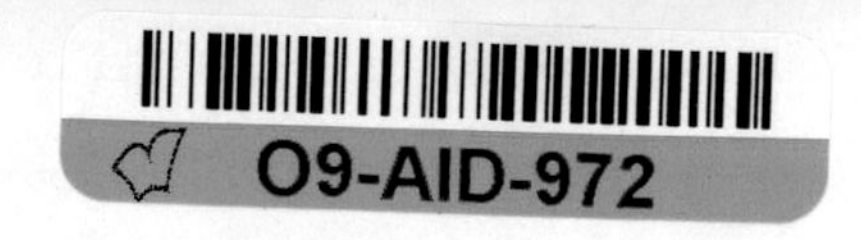
O9-AID-972

Programmable Logic Devices

Programmable Logic Devices

TECHNOLOGY AND APPLICATIONS

Geoff Bostock

McGraw-Hill Book Company
New York St. Louis San Francisco Colorado Springs
Oklahoma City San Juan

Library of Congress Cataloging-in-Publication Data

Bostock, Geoff.
Programmable logic devices.

Bibliography: p.
Includes index.
1. Programmable logic devices. I. Title.
TK7872.L64B67 1987 621.381 87-17230
ISBN 0-07-006611-6

1234567890 DOC/DOC 89210987

ISBN 0-07-006611-6

This book was first published in Great Britain in 1987 as *Programmable Logic Handbook* by Collins Professional and Technical Books, London.

Printed and bound by R.R. Donnelley & Sons Company.

This book is dedicated to Jenny, my wife,
who kept our home and business together
while it was being written.

Contents

Preface

Programmable logic devices have been available for over 15 years yet, until now, there has been no comprehensive reference work covering the subject. This is because it has been possible, until recently, for the design engineer to choose from a fairly restricted range of devices; restricted both in the sense of being small in number and simple to understand. In the last few years, however, the choice of architecture has become wider and more powerful in its application possibilities.

In compiling the data for this book it was clear that the starting point was hard to define; how much previous knowledge of semiconductors and electronics could I assume that the reader has? The easy way out was to assume that the answer was none. The early chapters are a summary of the basics of semiconductor devices and the principles of logic but, because this is intended as a reference work, there are no formal proofs of the results obtained. The references at the end of the book contain any formal working required by the reader. However tempting it is to miss out these early chapters, which may appear trivial to readers well versed in electronics, some of the later points have their origins in these basic teachings.

To put programmable logic into perspective let us take a brief historical look at the subject. PROMs were the first devices to come to the market, in about 1970, although whether they were intended as memories or logic is a moot point. The first true logic device was Signetics FPLA, introduced in 1974. Initially this was a limited success for a number of reasons; high complexity (for that time), inflexible architecture, large package and high price (compared with standard logic) being the chief ones. PALs, introduced by MMI in 1977, overcame many of the drawbacks of PLAs and took a lead in market share which they have never relinquished. The total market size is now over a billion dollars, shared between more than 100 device types.

Programmable logic has become a real competitor to standard logic on the one hand, and to masked ASICs on the other, but it still represents only a small fraction of their market size. There is thus a huge potential for growth with consequent benefits to designers, provided that they are in a position to take advantage of those benefits. This book is intended to help them take fullest advantage. All the currently available architectures are described in detail; the design methods are also covered. These range from manual techniques to high-powered CAE systems, showing that programmable logic is suitable for both low budget projects and highly equipped design centres.

One of the largest sections of the book is devoted to applications. These include simple logic functions which can be used to build more complex functions and some ideas for the more complex functions themselves. The applications are intended to be diverse enough to show how most logic requirements can be fitted into programmable logic devices. Even if the application is not covered exactly, enough information is provided to enable the designer to make the best choice of device and be guided as to how to complete the design.

The student with no prior knowledge of programmable devices should find this work a useful primer, particularly when used in conjunction with standard text-books covering the formal aspects of logic design. I hope that established designers will also find it helpful as a reference work to keep by their benches. If it helps to breed a generation of logic designers who turn to programmable devices before looking at lists of TTL and CMOS standard functions, then it will have achieved its goal.

Geoff Bostock

Chapter 1
Introduction to Logic Devices

1.1 BASIC PRINCIPLES

1.1.1 The idea of logic

In order to understand any discussion of what comprises a logic device it is necessary to be aware of what is meant by logic. In any system of logic, be it electrical or philosophical, the fundamental concept is that statements may be *true* or *false*. Conclusions about the state of the system being described are drawn from an analysis of which components of that system are true and which are false. For example, a simple combination lock might be devised in which two-way switches are placed in series with a relay. The relay would operate the lock mechanism; the system circuit is shown in Figure 1.1.

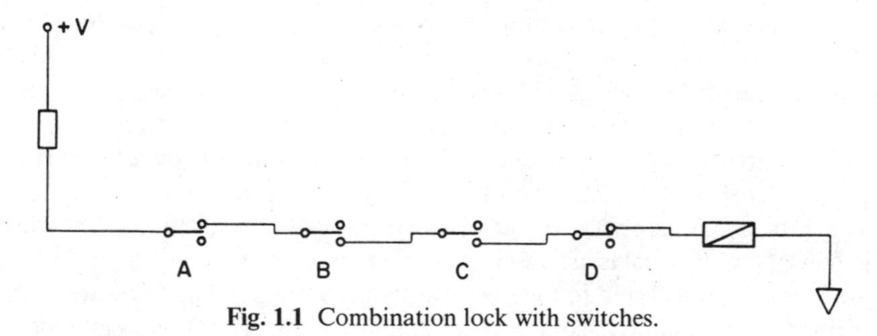

Fig. 1.1 Combination lock with switches.

Each switch is given a letter and only one combination of ups and downs will allow the lock to operate. In the system illustrated, with only four switches, it would not be very difficult to break the combination for there are only sixteen possibilities, but we will see in later chapters how the number can be extended to make a practical circuit. In our simple system it can be seen that the combination of A-up B-down C-down and D-up will open the lock. If up is equivalent to the *true* state and down is the *false* state, then a *logic equation* may be written to describe the circuit:

OPEN = A AND NOT B AND NOT C AND D

AND and NOT are the *logic operators* which define the relationships between the variables. In this example A and D must both be true and B and C false for the equation to be satisfied.

A similar system may be employed to make decisions about almost any situation. While the human brain is quite capable of making these decisions it usually needs the data to be converted to a visual form. Electronic logic devices take the data in the form of electrical signals and use electronic switches to implement the logic equations. In our example we used mechanical switches as the logic elements as well as the interface between the outside world and the electrical system. Usually the interface is separated from the logic elements and from now on we will concentrate on the electronic devices used to perform the logic. The most common electronic switch is the transistor and the next section describes the two types used in practical circuits.

1.1.2 Transistor switches

1.1.2.1 Semiconductors

Before describing transistor operation it is necessary to appreciate the materials from which they are made. Matter under normal conditions is composed of atoms. In solid matter the atoms are bonded together and held in relatively fixed positions by the interaction of the electrons in their outer layers. Conducting materials, such as metals, do not use all their electrons for bonding so the spares are free to move within the solid boundary and will conduct electricity. Other substances, particularly those with complex molecules, have no spare electrons and are therefore insulators. The effect of temperature is also relevant.

When heated a solid absorbs energy; internally this energy is stored as vibration energy by the atoms or molecules. In a conductor this has the effect of reducing the available space for the electrons to move around in, so the bulk resistivity of the material is increased. The effect on insulators is different. Some of the energy is transferred to the electrons, which are then able to escape from their bonding duties and become free to conduct electricity. Those materials in which this property is noticeable at room temperature, particularly monatomic crystalline solids such as silicon and germanium, are called semiconductors.

There is a more controllable mechanism by which semiconductors may be made to conduct electricity. A small amount of an impurity may be added to a crystal without disturbing the lattice too greatly, provided that the atoms of the impurity are similar in size to the parent atoms. If the impurity has more electrons than the parent available for bonding the spare electrons become available for conduction. This is called an *n-type* semiconductor because the current is carried by negative charges. Conversely, it is possible that an impurity will have fewer electrons available for bonding than the parent, in which case there will be *holes* formed in the bonding layer. Under the influence of an electric field, electrons will move to fill adjacent holes leaving a hole where they were; this makes it appear as if the holes themselves are moving through the crystal. Such material is called a *p-type* semiconductor as the current is carried by positive charges.

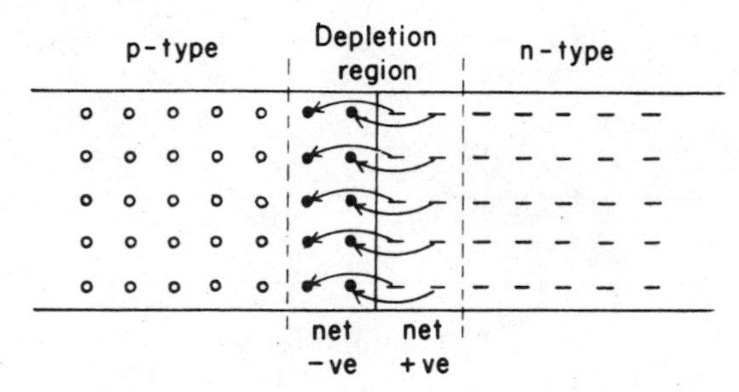

Fig. 1.2 p–n junction.

1.1.2.2 Diode junctions

While the current-carrying potential of a simple semiconductor depends simply on whether there is an excess of electrons or holes, a quite different situation exists when there is a junction between p-type and n-type material. At the junction itself there is a thin layer called the *depletion region* where the material is *intrinsic* and there are no free charge carriers. Figure 1.2 shows how the free electrons from the n-type side can diffuse to the other side and fill the holes, creating a potential barrier.

Applying a positive voltage to the n-type side pulls electrons away from the depletion region and increases the height of the barrier. In this case no current can flow through the junction. In the reverse case, when a negative voltage is applied to the n-type material, electrons are repelled towards the depletion region and cause the potential barrier to be lowered. When the barrier has been eliminated they meet holes which have been attracted from the p-type side of the junction. The electrons combine with the holes allowing a continuous flow of electrons in one direction and holes in the other. The net result is that there is a constant current flowing through the junction.

This property of the p–n junction, allowing current to flow in one direction but not the other, forms the basis of most electronic components from the diode to the VLSI integrated circuit.

1.1.2.3 MOS transistors

The MOS transistor is shown in cross-section in Figure 1.3. MOS is an acronym for Metal–Oxide–Silicon which describes the basic structure. The transistor is fabricated from a crystal of p-type silicon, into which impurities are diffused to form n-type regions called *sources* and *drains*. A thin layer of silicon dioxide is grown above the gap between each source and drain and a layer of metal or silicon deposited on the top. This top layer is called the *gate* and controls the current flow between the source and drain.

If the source and bulk silicon are held at the same voltage and the drain is taken to a more positive voltage then no current can flow between source and drain, because the drain-substrate junction is reverse-biased. If a positive voltage is now applied to the gate, electrons will be attracted into the region immediately

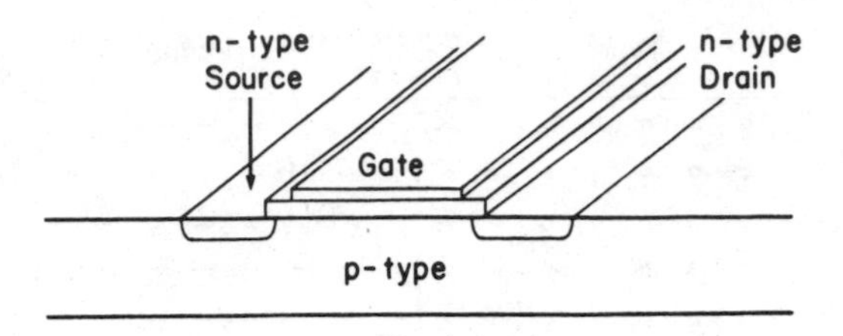

Fig. 1.3 MOS transistor structure.

below the oxide. This has the effect of making an n-type *channel*, which allows electrons to flow from source to drain. The voltage on the gate thus controls the flow of current through the transistor. A transistor of this type is called an *n-channel* device. A similar transistor made on an n-type substrate would be a *p-channel* device.

A MOS transistor can be used as a switch by connecting the current to be switched to source and drain, and connecting the controlling voltage to the gate. Because silicon dioxide is a good insulator, very little current has to be supplied by the control voltage; however, because the gate is acting as a capacitor there may be loading effects at high frequency. The channel is confined to a shallow region just below the surface and will therefore not permit very high currents to flow. The full consequences of these properties will be examined in a later chapter.

1.1.2.4 Bipolar transistors

If the n-type side of a p–n junction is doped more heavily with impurities than the p-type side, then many more electrons than holes will be attracted to the depletion region when the junction is *forward-biased.* Most of the electrons will then pass into the p-type material, where they will be 'eaten' gradually by holes. The electrons are said to be injected into the p-region. In a bipolar transistor, (Figure 1.4), the n-type region is called the *emitter* and the p-type region the *base*. The base is made very narrow and bordered by a second n-type region, the *collector*, which is usually made more positive than the base. Most of the electrons injected into the base will be attracted into the collector thus establishing a current flow between collector and emitter.

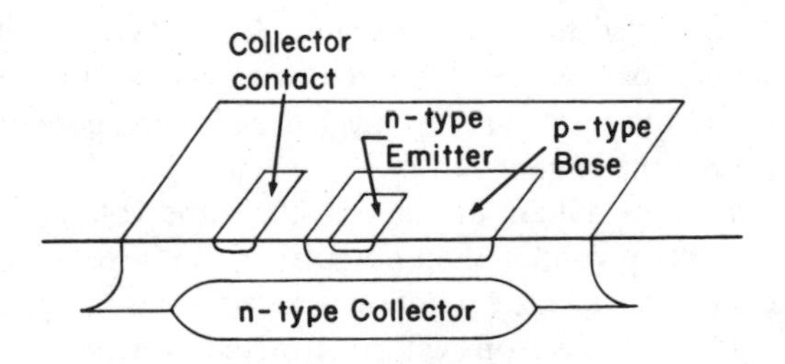

Fig. 1.4 Bipolar transistor structure.

The base controls the collector current as the voltage between base and emitter determines how much electron current is injected by the emitter. If the base voltage is too low to allow injection, then no current will flow in the transistor; thus the base acts as a control terminal for the bipolar transistor just as the gate does for the MOS device. Unlike the gate, the base must supply a small current to account for those electrons which are injected by the emitter but combine with holes before they reach the collector. However, because the area of the emitter can be made relatively large, a bipolar transistor can carry a larger current than an MOS.

The transistor described above is called an *npn* transistor to show the doping types of the three regions making up the device. It is also possible to make a *pnp* transistor where the dopings and voltages are reversed.

1.2 PRACTICAL LOGIC DEVICES

1.2.1 Planar technology

1.2.1.1 Masking and diffusion

Before embarking on a study of both standard logic families and programmable devices it is instructive to examine the technology used to fabricate integrated circuits. This is still based on the *planar process* developed in the late 1950s by the Fairchild Camera and Instrument Corporation. Figure 1.5 illustrates the steps required to create a p-type region in an n-type crystal of silicon. Slices, or wafers, about 1 mm thick are cut from a silicon crystal which has been grown by the *Czochralski* method. Current production uses wafers up to 150 mm in diameter. The wafers are chemically polished to remove mechanical damage incurred in the cutting process.

The first step is to grow a layer of silicon dioxide on the surface of the wafer. This is achieved by passing oxygen over the wafer in a furnace at a temperature of up to 1200° C. As many as fifty wafers may be processed at one time and furnace temperatures are controlled to better than 1° C. A thin layer of sensitive material is then applied to the surface of the wafer and the areas where p-type regions are required are defined. This may be achieved by exposure to ultraviolet light or an electron beam which polymerise the layer in areas which are to remain n-type. A photo-mask is used with ultraviolet light; the electron beam, which gives much finer definition, is electrically scanned over the wafer. The unpolymerised areas are dissolved in solvent to reveal the silicon dioxide surface, which is then chemically etched by hydrofluoric acid to expose the underlying silicon.

The next step involves another high-temperature furnace operation; in this case a gas containing the required impurity is passed over the wafers and forms a solid solution at the surface of the silicon. Prolonged exposure to temperature causes the impurity to diffuse into the silicon to a depth of a micrometre or more. The net result is a tub of p-type material in the n-type; the boundary between the

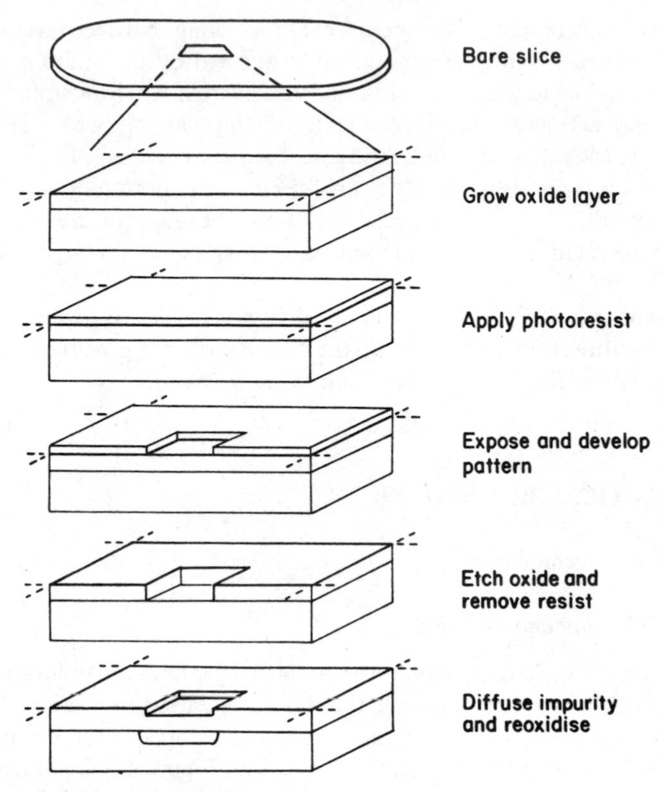

Fig. 1.5 Masking and diffusion steps (planar process).

two being a p–n junction. Successive diffusion steps are required to build up the transistor structures described in the previous section.

1.2.1.2 Metallisation

Having fabricated silicon-based components a way has to be found to enable them to be used. This involves connecting some kind of rigid metallic structure to the silicon to enable the device to be mounted, for example, on to a printed circuit board. The size of transistor features, usually a few micrometres, means that direct connection of wires is virtually impossible for reliable permanent joints. In practice, photolithography has to be used again. A metal, such as aluminium, is evaporated over the silicon surface to form a thin film. Windows previously etched in the silicon dioxide allow the aluminium to make contact to the silicon where connection is required. The aluminium itself is then etched to form conductive tracks from the silicon windows to metallic areas which are large enough to allow direct connection of wires.

Where the aluminium contacts the silicon an alloy is formed to ensure reliable connection. Aluminium is a p-type impurity in silicon so some care has to be taken when connecting to n-type areas. If the n-type area is heavily doped the depletion region will be extremely thin and holes and electrons will cross it very easily even when it is reverse-biased. There will be no problem in connecting aluminum to p-type silicon as no junction is formed in this case.

1.2.1.3 Integrated circuits

As well as providing a conducting path from the silicon to the connecting wires, aluminium tracks can be used to connect diffused components in the same silicon wafer. Circuits containing several transistors, diodes, resistors and even capacitors can be connected-up on the silicon wafer surface, just as the discrete components can be connected on a printed circuit board. In principle, there is no limit, within the ingenuity of the designer, to the number of components which can be connected on a wafer. In practice, the planar process is subject to random faults caused by dirt particles or material defects. If a component has a fault it will not function correctly, so the circuit containing it will also be defective.

The potential number of circuits on a wafer depends on the area of the wafer and the area of a circuit. Probability theory can predict what proportion of these circuits will be faulty for a given level of fault densities. This enables manufacturers to calculate the largest circuit which they can make economically. Improvements in processing and materials technology cause a steady reduction in the fault density allowing larger circuits to be designed and manufactured.

1.2.1.4 Packaging

The final stage in integrated circuit manufacture is packaging. Most electronic assemblies are based on printed circuit technology so the integrated circuits have to be packaged into a form which is robust enough to withstand the handling they will receive, while allowing the connecting pins to be soldered into the *Printed Circuit Board* (PCB). The standard package is the *dual-in-line* which has two rows of pins fitting into a 0.1 in. grid. The pins are 0.1 in. apart; the row separation depends on the number of pins, usually 0.3 in., 0.6 in. or 0.9 in.

The packages are constructed on a frame, called the *lead frame*; the circuit itself is alloyed to a central bar and fine wires, about 25 μm thick, bonded from the circuit to the pins. The whole frame is either sandwiched between two ceramic plates in an inert atmosphere, or moulded into solid plastic. The circuit is cropped out of the supporting frame and the leads formed into the conventional inverted 'U' to allow easy insertion into holes in the printed circuit board. Package sizes from eight to sixty-four pins are found in this form; the number of pins and package width are related by the mechanical problems of fitting the leads round the circuit. Reducing package size for a given pin-count is a major task, which has received the attention of most integrated circuits manufacturers.

A recent development is the *SO package* which has leads on a 0.05 in. spacing formed into an 'L' shape, which allows the circuit to sit on the board, instead of the pins passing through holes. The package may be stuck to the board and connection made by means of a solder paste laid down previously by a screen printing process. The smaller spacing means smaller packages, and other components are available in similar styles. Component placing may be mechanised so the production process becomes cheaper, and the size reduction also means cheaper mechanical components. The whole process is called *surface mounting* and is probably the way that most production will be done in the future.

As logic functions become more complex they often need more connections, and the DIL and SO packages cannot cope with the pin numbers required. Another type of surface mounting package has been developed, based on having leads on all four sides. The first packages of this kind had a ceramic base with printed connections leading to plated areas at the edge of the package. These were intended for direct soldering to a ceramic substrate, as found in thin and thick film circuits. Special sockets were developed to enable these packages to be used on printed circuit boards; their lack of leads led to them being called *leadless chip carriers*, LLCCs for short. Plastic versions in the same style use leads bent into a 'J' and located in notches at the edge of the package; they are called *Plastic Leaded Chip Carriers*, or PLCCs. PLCCs and LLCCs have been developed for packages as small as twenty leads, but they are more useful for circuits with much larger counts, from forty-four to over 150 now possible.

1.2.2 Standard integrated circuit families

1.2.2.1 Bipolar logic circuits

Bipolar transistor action depends on the properties of p–n junctions which are below the silicon surface, albeit by only 1 or 2 μm. The active region of a MOS transistor is at the surface itself. The surface properties of silicon, and the ways to control them, were understood at a later time than the bulk properties, so bipolar logic circuits evolved before MOS. Figure 1.6 shows how bipolar silicon components can be connected to form an AND circuit.

If all three switches are open then no current will flow in any of the three input diodes. The base of the output transistor is thus connected to a positive voltage, so the transistor is switched on, current flows in the output resistor and the lamp will be turned on. If any of the switches is closed current will flow through the diode connected to it, so the voltage on the transistor base will be insufficient to

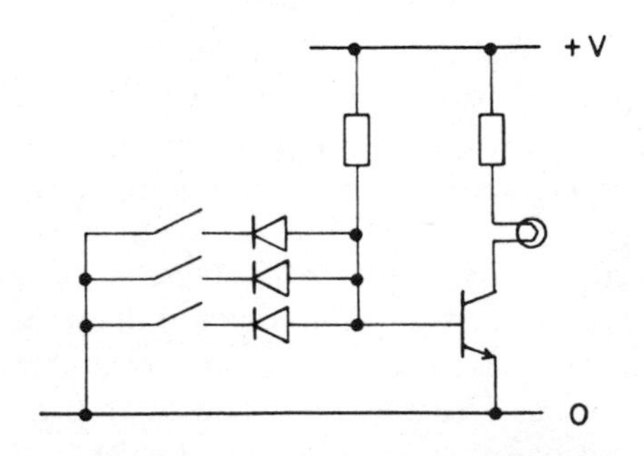

Fig. 1.6 AND function – bipolar.

allow it to conduct appreciably, and the lamp will not be lit. The lamp turns on only when all three switches are open so the circuit implements the AND function.

This circuit as it stands is not quite suitable to form the basis of a logic family. By amalgamating the input diodes into a single multi-emitter transistor and adding an output stage which will drive succeeding circuits the basic *transistor–transistor–logic*, or TTL, circuit of Figure 1.7, is formed. A whole family of logic functions based on the simple circuit described above has been designed over the past twenty years and now forms a major part of the circuit designer's armoury.

1.2.2.2 MOS logic circuits

Once the properties of silicon surfaces were understood, and the techniques for controlling them had been perfected, it became possible to construct logic

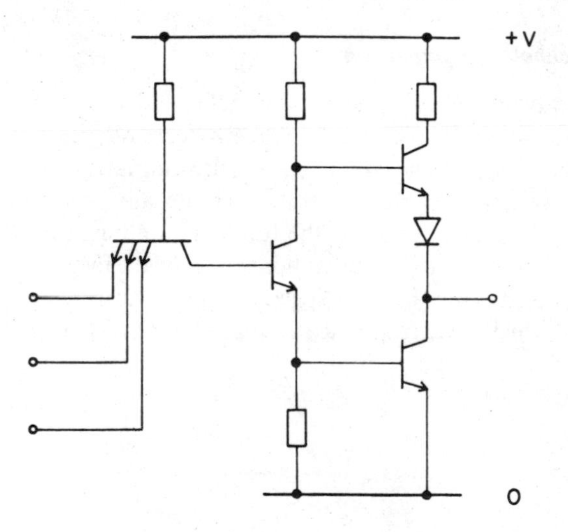

Fig. 1.7 TTL NAND gate.

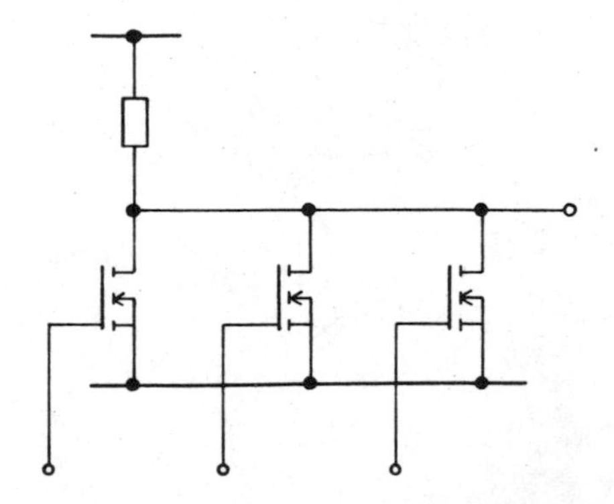

Fig. 1.8 NOR function-n-channel MOS.

circuits using MOS switches. The circuit in Figure 1.8 shows how a simple NOR gate can be built from MOS transistors. The transistors are *n-channel* type so any gate taken to a positive voltage will open a conductive path from the load resistor to ground. This structure will, however, provide no particular advantage over bipolar transistor circuits; indeed, the performance is likely to be worse because of the lower current capability and higher capacitance of MOS.

The identical function can be constructed from *p-channel* transistors, as shown in Figure 1.9, by connecting them in series. Any gate taken to a positive voltage will turn its channel off and prevent current from flowing in the chain. Again, no advantage is obtained from this structure, but consider what is achieved by combining the two, as in Figure 1.10. If any inputs are taken positive the *p-channel* transistors are switched off, while if all the inputs are negative the *n-channel* transistors do not conduct. Thus, no current flows in any steady state condition. The result is a logic circuit taking very little current; there is now a decided advantage over bipolar transistors.

1.2.2.3 Technology comparison

Some of the advantages of bipolar against MOS and vice versa have already been mentioned above. The choice between the two technologies is essentially dictated by practical considerations which are worth summarising here. The two chief performance criteria under consideration are usually speed and power. That is how fast the circuit will perform the function contained in it, and how much power must be supplied to it in operation. Usually, the faster the circuit the higher the power consumption, because speed is achieved by charging circuit capacitance quickly which involves the use of higher electrical currents.

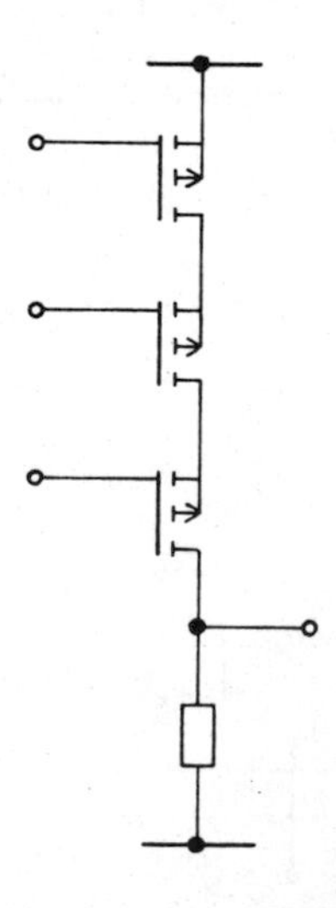

Fig. 1.9 NOR function – p-channel MOS.

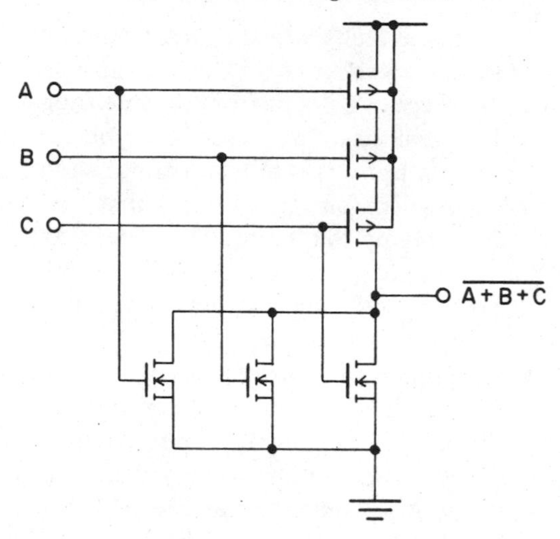

Fig. 1.10 NOR function – CMOS.

As already noted, bipolar transistors are better at handling high currents and so bipolar technologies usually offer higher speed. Examination of the technology shows, however, that MOS components can be diffused closer together and intrinsically smaller. This is because bipolar transistors need to be isolated from each other while groups of MOS can be diffused together. Also the diffusion windows in bipolars need aligning with each other while many of the MOS processes arc sclf-aligning. MOS structures therefore have lower capacitance, and can operate at lower currents for a given speed. This is no real advantage where simple circuits are concerned because the major component of capacitance is in the external circuit which is connected to it. As circuits become more complex the internal capacitance becomes more important and MOS becomes a better choice for high speed.

The MOS circuit described above using both p- and n-channel transistors is termed *Complementary MOS*, or CMOS for short. The other main advantages of CMOS are the higher noise immunity inherent in the gate design – CMOS switches virtually between the two power rail voltage levels – and more complex functions can be built into the same area of silicon because of the higher packing density of MOS transistors. On the bipolar side are the capacity to make the absolutely fastest circuits, at the cost of high power consumption, and greater robustness, because MOS transistors are susceptible to relatively low static electricity voltages on their input gate oxide.

1.3 CUSTOM LOGIC CIRCUITS

1.3.1 Microprocessors

The first integrated circuits were only simple functions capable of containing a single logic equation. Attempts to make larger circuits failed because the silicon processing introduced random faults, which limited the area of silicon usable by a single circuit. This was acceptable to equipment designers, who were used to thinking in terms of what could be built from discrete transistors on a single printed circuit card. The replacement of a dozen components by a single metal can containing the same components in miniature form was an obvious benefit to them. Improvements in silicon processing meant that more complex circuits could be integrated, but only up to a point. There are certain logic functions which are used universally in system design and it clearly makes sense to integrate these, but above this point the way that these functions are combined becomes specialised or *application-specific*.

One way in which the component industry reacted to this situation was a result of its close involvement with the computer industry. Computer manufacturers were the strongest influence on component designs because the performance of computers was strongly dependent on the speed and complexity of the circuits from which they were built. A computer is a highly complex logic system capable of implementing logic and arithmetic equations by virtue of a set of instructions which modify the internal logic functions according to the instruction being carried out. There is no reason in principle why a computer cannot be built on a single silicon circuit and this was the route taken by some component manufacturers. Because of their small size, and because the number of functions they could manage was limited, the name *microprocessor* was coined for these devices. The function of the microprocessor was determined by the sequence of instructions they were controlled by, rather than their actual circuit layout, so they were complex circuits which could be made application-specific by the equipment designer.

1.3.2 Custom circuits

The main reason for the upper limit on complexity of standard integrated circuit functions is concerned with the economics of integrated circuit manufacture. The cost of designing even a simple circuit and getting it into production is more than £100000 ($140000), so manufacturers rely on selling millions of circuits to recover the design costs. In some cases, however, an equipment manufacturer may feel that it is worth while to invest that sum to obtain the benefits of having a customised circuit built for them. The equipment maker will save on his production costs by replacing several circuits with one and will have the added advantage of secrecy, since it would be difficult for a competitor to copy his system without knowledge of the custom circuit.

This course of action will only be possible if a substantial number of customised units can be built – probably well in excess of 10000. Integrated circuit manufacture is a batch process in which over a thousand good circuits should result from a single batch, even for complex designs. Because of the other setting-up costs, such as mask making and test program writing, most manufacturers would be unwilling to commit these resources to just a single batch of custom circuits. The full-custom option, then, is only viable for well-proven designs committed to long production runs. Another problem for the unwary is that it is not always straightforward to transfer a design involving several discrete circuits on to a single silicon circuit. Typical problems are caused by timing changes and layout interactions which may result in several iterations of the mask set. Good computer simulation of the circuit can cut down the problems; however, a job which may start out with an expectation of completion in twelve months, long enough for a circuit design anyway, may finish up taking twice as long and costing twice as much.

1.3.3 Gate arrays

Many manufacturers have introduced the *gate array* concept which is a good compromise between the standard and custom circuits. A gate array has a base layer containing hundreds or even thousands of identical simple logic functions diffused in a regular pattern, or array, in the silicon. The whole design is usually very software-intensive with a library of standard functions called *macros* which can be built from the basic functions. The designer defines the way in which the macros themselves are interconnected, and computer programs then physically define the layout on the array and simulate the electrical performance based on this layout. Iteration to obtain the desired performance is relatively quick and the program will usually produce the interconnection masks and test programs automatically. It now becomes economic to produce only a thousand circuits of a given design and allows smaller users to enjoy the benefits of customisation.

Being a compromise, this approach does not usually yield as high a performance as full-custom solutions, nor is it as flexible in the range of circuit complexities which can be accommodated. Other approaches use a variation of the gate array topology by offering, for example, standard cells, completely flexible gate counts or direct writing by electron beam to improve performance or speed up prototype deliveries. In all these cases there are a number of common factors:

- a commitment to some minimum quantity
- a development charge payable to the manufacturer
- the manufacturer controls the final production

Thus the consumer who wants just a few, or is unwilling to commit up-front to a full design or needs a fast delivery response, is unlikely to be an array or custom circuit user.

Chapter 2

Programmable Device Techniques

2.1 MEMORY DEVICES

2.1.1 ROM structure

It was stated in the previous chapter that microprocessors are controlled by a sequence of instructions which determine the internal function at any time. Along with the microprocessor, a method of storing and reading the instruction sequence had to be developed. The obvious way of ordering the instruction sequence is to label the first instruction '1', the second '2' and so on. The instructions can then be stored in some device which will respond to the number '1' with the first instruction; when this has been done the microprocessor can send it number '2' and it will respond with the second instruction and continue in this way until the operation is complete. Conventionally the number of the instruction is called its *address* and it is usually counted in hexadecimal notation, because microprocessors are almost all organised in multiples of four-bits, or *nibbles*.

One form of microprocessor architecture is shown in Figure 2.1. This is called Harvard Architecture and is rather wasteful of connections, but it is very easy to understand. There are three groups of connections to the microprocessor, each group being called a *bus*. The *data bus* allows the microprocessor to communicate with the outside world, the *address bus* is the output for the next address and the *instruction bus* is the route back into the microprocessor for the next instruction. We need not concern ourselves with the internal workings of the microprocessor, but rather with the box labelled 'ROM' which takes the address and sends back the instruction.

ROM stands for *Read Only Memory* which implies that it is fixed and cannot be changed. The ROM must recognise the address which it is receiving and find the instruction corresponding to that address. For simplicity, let us assume that the address bus is a single nibble so that the ROM can recognise an input in the range 0–F. Each single input line can be at a *high* voltage or a *low* voltage corresponding to logic *true* or logic *false*, or to binary number '1' or '0'. Thus, when the microprocessor wants instruction '2' from the ROM it will send out '0010' on its address bus. Inside the ROM are sixteen detectors, each responding to a different combination of *highs* and *lows* on its address input lines. As Figure 2.2 shows, each detector is connected to a set of switches to send either the *high* from the detector or a *low* to the appropriate output. This is the instruction code which is sent back to the microprocessor.

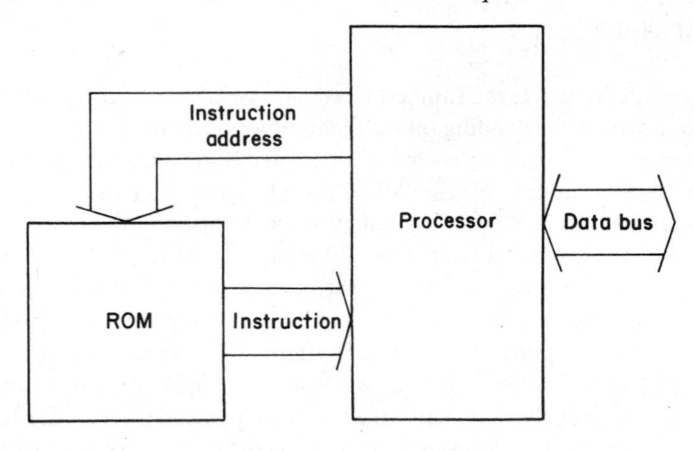

Fig. 2.1 Harvard Architecture.

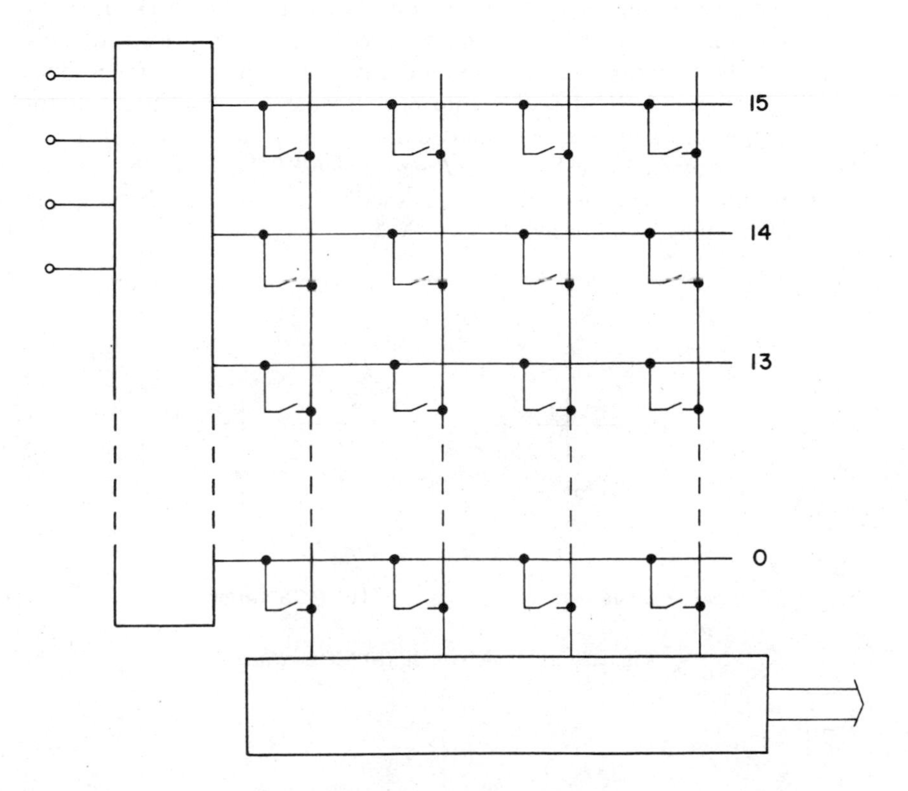

Fig. 2.2 Basic ROM structure.

The electronic method of decoding the input signals is a function to be found in the standard logic families, described in Chapter 1 and explained in more detail in Chapter 3. For the rest of this chapter we will describe the various ways in which the switches can be physically incorporated into ROMs.

2.1.2 Masked ROMs

As we saw in Chapter 1, the simplest electronic switches are either the diode or the MOS transistor, depending on the technology being used. Figure 2.3 shows how either may be used to switch the detected signal; in fact, a number of different configurations of the MOS switch have been used by various manufacturers, but the principle remains the same. On the silicon surface, contact between the metal interconnections and the active devices is made via contact holes in the silicon dioxide covering the surface. Thus the connection may be permanently made or broken, according to the presence or absence of a contact hole to the appropriate switch. In Figure 2.4 contact is made to the left-hand diode and the output will be *high* when the address corresponding to that switch is input, while the right-hand one remains open and the output will be *low*.

The information in the ROM is thus stored as the presence or absence of a hole in the silicon dioxide, it being physically programmed into the device during manufacture, according to the information coded into the mask used for opening the contact windows. As we noted earlier, masking processes are only viable for large quantities and take several days, or even weeks, to produce devices. The masked ROM is therefore not suitable for small users or for prototyping, when changes may be needed and tested at short notice. This has led to the development of the PROM or *Programmable ROM* which the user can programme himself. The following section describes the various technologies in which PROMs may be fabricated.

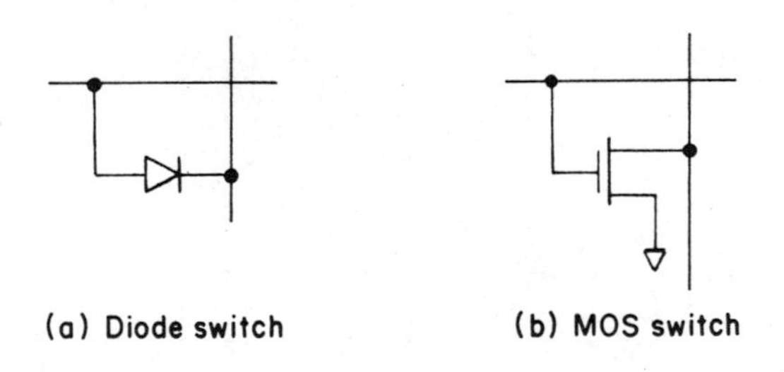

Fig. 2.3 (a) Diode switch; (b) MOS switch.

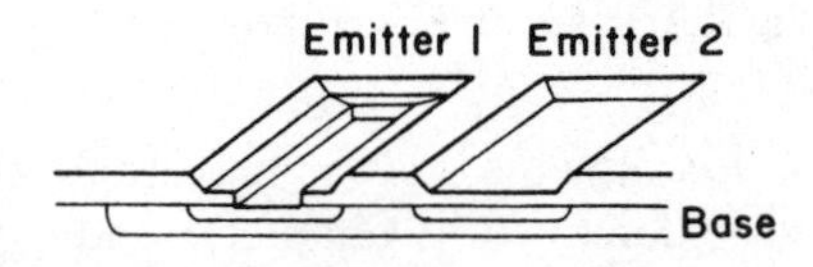

Fig. 2.4 Masked ROM programming.

2.2 FIELD-PROGRAMMABLE ROMS

2.2.1 Metal fuses

2.2.1.1 Fuse structure

In order to make a ROM *field-programmable* some way must be found to replace the holes in the contact mask with a switch which can be opened from outside the device. Perhaps the simplest switch is a metal fuse, whose operation in a domestic context is quite familiar. A thin piece of wire is placed in the current path; when the current exceeds a given value the heating effect is sufficient to melt the wire and switch off the current. The switch may only be closed again by replacing the wire, so it is an effective safety device.

On the microelectronic scale, electrical connection is achieved by thin films of metal, rather than wire, and a fuse may be fabricated in the same way. As discussed in Chapter 1, aluminium is the usual choice for interconnecting metal, the reasons being that it is easy to evaporate into a reliable thin film, and it has a relatively low resistivity. The requirements for a fuse metal are not quite the same. It must be capable of being evaporated into a thin film, but a high resistivity is preferred as the heat produced by an electric current is proportional to the resistance. Common materials are alloys formed from nickel and chromium, or from tungsten and titanium, while some manufacturers use polycrystalline silicon itself or platinum silicide. A number of different shapes of fuse are used, the commonest being shown in Figure 2.5 as the bar, taper and notch. The advantages of each will be discussed in the Section 2.3.3 on quality and reliability. A common feature of each is the size, being typically about 5 μm wide and 10 μm long.

2.2.1.2 Programming method

Although it is relatively straightforward to fabricate an array of metal fuses and diodes, some way has to be found to blow out the unwanted fuses without damaging any of the other components on the circuit. The typical fusing current is 50–100 mA which could destroy the small signal components within the circuit. Figure 2.6 shows one element in the fuse array with the diode replaced by a transistor; diodes in integrated circuits are formed from transistors in any case.

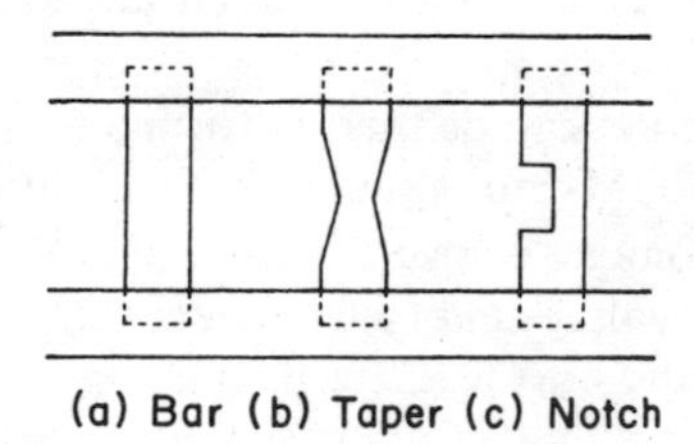

Fig. 2.5 Fuse shapes. (a) bar; (b) taper; (c) notch.

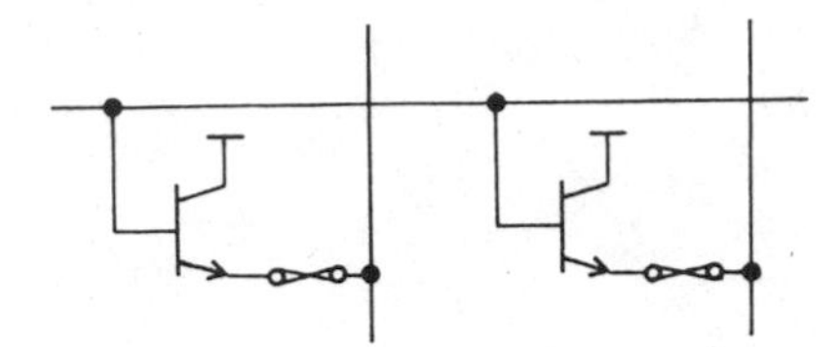

Fig. 2.6 Transistor switch array.

The base of the transistor is driven by the address selection circuit, while the emitter drives a buffer which forms one bit of the output word. The collector is connected to the supply voltage (V_{cc}) so that the drive to the output buffer does not have to be supplied by the address selection circuit. If this transistor is made large enough it could supply the programming current for the fuse, but the output buffer cannot sink this current for its input is usually just the base of a transistor. Besides, there has to be a way of defining that the fuse is being programmed, as opposed to normal circuit operation when it must be made impossible to blow the fuse.

The way round this is to use another property of diodes. Although applying a reverse voltage to a junction increases the potential barrier which charge carriers outside have to climb, it also increases the electric field inside the depletion region itself. Inside the depletion region the silicon behaves as if it were intrinsic, or undoped. In this situation thermal agitation causes a few charge carriers to be produced spontaneously. The electric field sweeps them out of the depletion region and they appear as *leakage current* to the outside world. At a critical value of electric field they acquire enough energy to create further carriers in collisions during their passage through the depletion region. This multiplication causes a rapid increase of leakage current and is termed *avalanche breakdown* of the junction. It occurs at a voltage which depends on the doping levels on either side of the junction, and is called the *breakdown voltage.*

A rather different mechanism can occur in very heavily doped junctions; this is called *Zener breakdown.* In this case the mechanism is somewhat different and results from heavily doped junctions having a very narrow depletion region. Carriers with sufficient energy can *tunnel* through the depletion region; this is a process which can be described mathematically by quantum mechanics. Physically, what happens is that the electric field breaks the chemical bonds holding the atoms together and allows current to flow through the depletion region.

The emitter-base junction can be used to form a well defined *zener diode* and is used in bipolar PROMs to enable the fuse-blowing circuitry. Fusing specifications call for voltages above the normal 5 V to be applied to V_{cc} and output pins. The V_{cc} overvoltage may select larger transistors to supply the fusing current; although the transistors used for normal operation could be made large enough this would cause a speed penalty, so it is usual to have a second set of drivers for fusing only. The output overvoltage will cause the output buffers to be bypassed and enable a set of transistors which can sink the fusing current. The

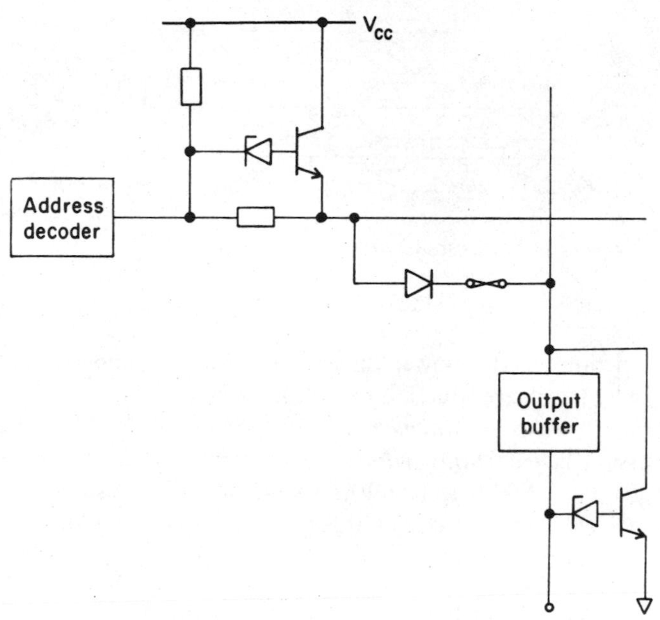

Fig. 2.7 Basic programming circuit.

fundamental circuit principle is shown in Figure 2.7; actual circuits used may differ in detail as various techniques have been devised to improve programming performance.

2.2.2 Diode fuses

2.2.2.1 Fuse structure

Although metal fuses are well established – indeed, they have been used in production devices since 1970 – they do increase substantially the size of the device compared with an equivalent mask-programmable circuit. The fuse itself is an extra component on the silicon surface associated with each diode in the array, while the mask programming is built into the diode structure itself. The diode fuse overcomes this drawback by making the programmable element part of the diode itself.

Each cell consists of a diffused transistor with the base connection left open-circuit; see Figure 2.8. As described in the previous section, a reverse voltage in excess of the breakdown voltage will cause a large current to flow through the junction. This property is used in zener diodes when the current flow is restricted to limit power dissipation in the junction, thereby avoiding any damage to it. In a diode fuse sufficient current is allowed to flow for the junction to melt locally; aluminium from the surface is carried by the current flow into the junction, where it alloys with the silicon. The alloy thus formed has a lower resistance than

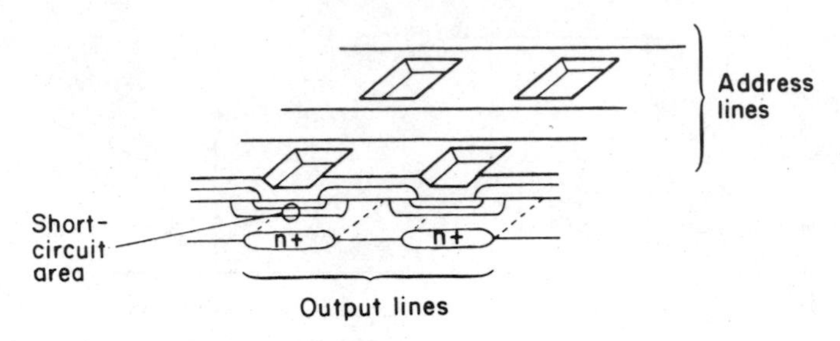

Fig. 2.8 Diode fuse.

the silicon junction, so the power dissipation decreases and damage to the junction is limited to the production of a short circuit.

The process is known as *Avalanche Induced Migration*, or AIM; there is an enhanced version called *Diffused Eutectic Aluminium Process*, or DEAP (a trademark of Fujitsu Microelectronik), in which additional silicon and metal layers are deposited over the diode in order to control the alloy formation more closely.

2.2.2.2 Programming method

As with metal fuses, the fusing circuitry is enabled by overvoltages feeding through zener diodes. Unlike metal fuses, current flows after the fuse has been blown. The time taken for either metal fuses or diode fuses to blow is not exactly definable. It is normal for the programming time for a metal fuse to be set much longer than the actual time taken for it to fuse. No current flows after the fuse has blown, so no additional stress is suffered by the fusing circuitry. If this procedure was adopted for diode fuses, the components in the fusing path would be carrying the fusing current for much longer than necessary. Thus, the programming method calls for a number of short pulses, with a verification after each pulse to see if it is possible to end the sequence. Once a short circuit has been detected it is usual to apply a final pulse to burn it in.

With so much heat being dissipated and so much current flowing it might be expected that the collector–base junction would be damaged also. The collector–base junction has a larger area than the emitter–base so the heat generated there will not cause such a large temperature rise; furthermore, the voltage drop across the collector–base is an order of magnitude lower because it is forward-biased. Nevertheless, one of the most important considerations in designing the programming specification is to ensure that the remaining diode is not degraded when the diode fuse is short-circuited.

2.2.3 MOS floating gate cell

2.2.3.1 Floating gate structure

Both the metal fuse and the diode fuse share the need for a large current pulse to blow the fuse and, hence, program the PROM. As we noted in Chapter 1, only the bipolar process lends itself readily to the passage of high currents. Fuse link and AIM PROMs are almost exclusively made by that process; the one exception is where silicon itself is used as the fuse material. Some CMOS PROMs are made with silicon fuses because a higher resistivity is achievable than with metal, and therefore lower fusing currents can be used, but the resistivity is harder to control so this technology has been less successful.

A completely different memory cell has been developed for MOS devices based on the operation of the MOS transistor itself. The structure, called a *floating gate*, is illustrated in Figure 2.9. There are two gates, the upper one is the normal gate connected to the output from the address-decoding circuit. The lower gate, on the other hand, is isolated electrically from it and from the channel. The action of this floating gate is to modify the threshold voltage of the MOS transistor, depending on whether or not it is charged. Charging takes place by injection of high-energy electrons on to the gate through the surrounding oxide; the oxide prevents the charge from escaping during normal operation. Figure 2.10 shows the effect of the two threshold voltages on conduction through the cell. When the floating gate is charged it raises the threshold to a level such that the transistor cannot conduct when 'normal' voltages are present on its gate. Thus the MOS transistor acts as a switch which can normally be turned on and off by the address selection circuit, but is permanently off after programming.

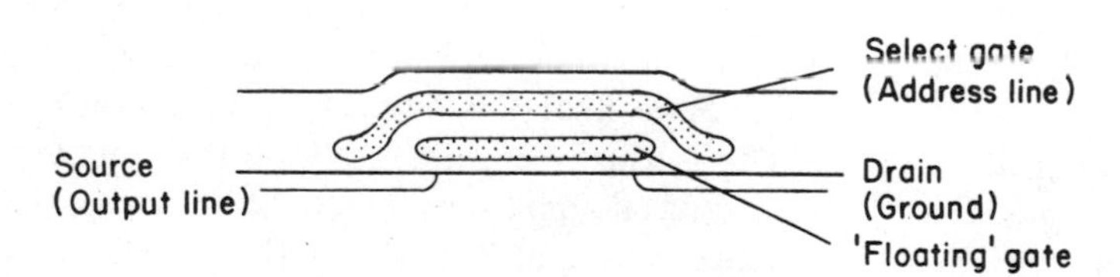

Fig. 2.9 MOS floating gate cell.

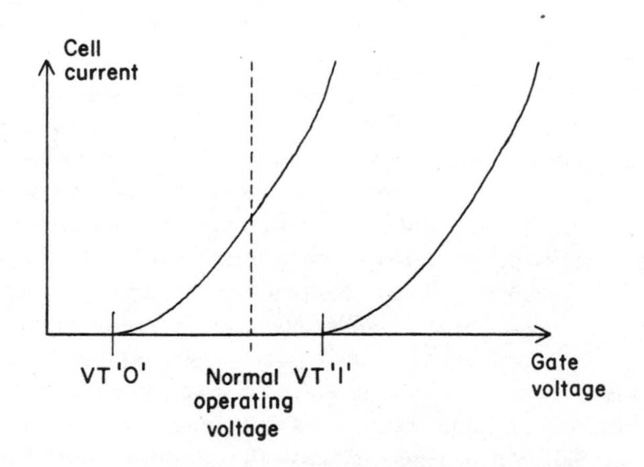

Fig. 2.10 MOS cell thresholds.

2.2.3.2 *Programming methods*

As stated above, the floating gate is charged by injecting high-energy electrons through the surrounding oxide. The oxide layer between the two gates acts as a potential barrier to electrons, but may be overcome by applying a high voltage (26 V originally) between the control gate and the silicon substrate. Although there is no direct connection to the floating gate, it will acquire some intermediate potential which remains when the voltage source is removed. The cell to be programmed is selected by addressing it through the address decoder and the output pins. Some MOS PROMs incorporate a separate pin which enables the addressing circuits in the programming mode.

The method of programming has developed since the introduction of MOS PROMs. The first devices called for the charge to be built up gradually by cycling through the addresses with a short (<1 ms) pulse at each address. This minimised the problems caused by heating and capacitative coupling between adjacent cells. Improvements in design and processing allowed this early method to be replaced by what is now the standard method, that is, to use a single 50 ms pulse on each cell. This leads to an inconveniently long time to program a single PROM; nearly 7 min. for a PROM with 8192 addresses. To improve matters so-called 'intelligent' algorithms have been developed.

Intelligent methods rely on using a series of short pulses which are counted by the programming equipment until the cell is detected as being programmed. Usually a 6 V supply is used to increase the threshold voltage which must be overcome. Once the cell has become charged enough to register as being programmed, a final programming pulse is applied to ensure that the floating gate is fully charged. The length of the final pulse is calculated from the number of short pulses first applied. This is usually 2–3 times quicker than the standard 50 ms pulse, which has to be set long enough to ensure reliable programming for all cells in the PROM. In practice, if several PROMs are being given the same program a *gang programmer*, which can accommodate eight or sixteen PROMs, is used.

2.2.3.3 *Erasing MOS PROMs*

Once a metal fuse or diode fuse has been blown it is impossible to reverse it to make the connection again. The floating gate is a different case, however, since no irreversible process is involved in charging the cell. At first sight, the obvious way to reverse the process is to apply a negative voltage across the floating gate, in order to discharge it. The result of doing this might be to finish up with the gate charged negatively, which would leave the threshold voltage negative; this would allow current to flow irrespective of the positive voltage applied to the control gate. In any case, it would be impossible to determine exactly the end point of the discharging process to arrive at a satisfactory unprogrammed state.

A way must be found to allow the stored charge to be removed completely without introducing any spurious charges or damaging the cell. If electromagnetic radiation falls on a semiconductor the electrons will absorb energy, provided that the energy of each quantum is enough to allow the electron to free itself from the bound state in which it is held. This principle is used in photodiodes where energy absorbed by a p–n junction produces electron-hole pairs which, in turn, create a large increase in the junction leakage current. Infrared quanta are sufficiently energetic to excite a photodiode, but the electrons in the floating gate have a much higher potential barrier to cross. The MOS cell needs ultraviolet radiation to supply sufficient energy to the stored charge for it to cross the surrounding oxide.

MOS PROMs supplied in a package with a window to allow the data to be erased by ultraviolet radiation are called *Erasable PROM*s or EPROMs. Care has to be taken to give EPROMs sufficient time under the ultraviolet otherwise the floating gates will not be completely discharged; then the resulting threshold will be indeterminate and cause incorrect data to be sensed. EPROMs have a great advantage over fuse-link PROMs when developing programs for microprocessors. It is very common to want to make changes to programs, or to try small parts of them, in which case the ability to reuse PROMs makes program development much less costly than would otherwise be the case.

2.2.4 Electrically erasable PROMs

2.2.4.1 Structure

Figure 2.11 shows that the electrically erasable cell is very similar to the floating gate cell. Instead of the floating gate, it uses charge stored in surface state sites at the interface between a layer of silicon nitride and silicon dioxide. Wherever there is an interface between different materials the disruption caused to the crystal lattice on an atomic scale leaves sites which can be filled by free charges passing through the interface. As with floating gates, there is a large potential barrier for charges to overcome if they are to travel from the control gate to the interface sites. This cell is commonly called a *Metal Nitride Oxide Silicon*, or MNOS, cell to underline its metal nitride oxide silicon structure.

2.2.4.2 Programming and erasing method

MNOS cells are programmed in much the same way as floating gate MOS cells, that is, by applying a short voltage pulse between the control gate and substrate of the cell being charged. Electrons are attracted into the empty interface sites and raise the threshold voltage of the transistor. Because it is surrounded by silicon nitride and silicon dioxide, which are both good insulators, the charge stays in place until it is forced to move.

The difference from the floating gate cell comes in the erasing method. There are only a limited number of interface sites and they can only accept charges of one polarity, so it now becomes feasible to remove the charge by applying a reverse voltage to the control gate. Once the sites have been emptied no more charge can move and the cell is in the unprogrammed state. Apart from being, arguably, more convenient to erase the device electrically rather than by ultraviolet irradiation, it means that the cells can be erased an address at a time

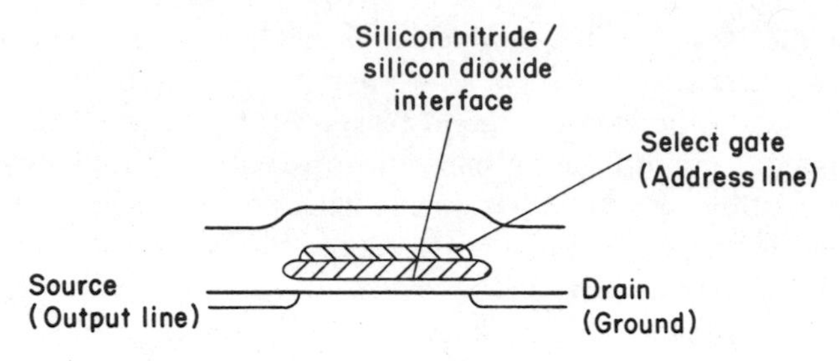

Fig. 2.11 MNOS erasable cell.

so that complete erasing and reprogramming is not necessary whenever a change to the stored data is needed.

2.3 PRACTICAL CONSIDERATIONS

2.3.1 Bipolar PROM designs

2.3.1.1 The diode array

Commercially available bipolar PROMs range in size from 32 × 8 (five inputs, eight outputs) and 256 × 4 (eight inputs, four outputs) to 8192 × 8 (thirteen inputs, eight outputs). Design of the diode array affects both the performance of the PROM and, because of the effect on chip size, the ultimate cost of the device. Clearly, it must be possible also to supply enough current to blow the fuses in the finished structure. Moreover, the problems involved in designing a small PROM are likely to be a lot less than those for a large PROM. Figure 2.12 shows the actual structure of a 32 × 8 PROM.

Depending on the pattern of HIGH and LOW voltages on the five input lines, one of the horizontal lines, or *rows*, will be set to a low voltage. Those crossovers which still have an intact fuse will take the corresponding vertical line, or *column*, to a low voltage through the diode at the crossover. All the other diodes on that column are reverse-biased, so none of the other rows will be affected by the columns which have been pulled LOW. The pattern of HIGHs and LOWs is taken to the outputs via the output buffers, which provide enough current drive to interface with the outside world.

This pattern may not be extended indefinitely in order to make a much larger PROM. For example, an 8192 × 8 PROM would require 8192 rows and eight columns which would make for a very difficult connection problem on the surface of the PROM chip. It would also cause the area of the chip, and its power consumption, to be very large because each row signal has to be derived from the address lines by some kind of logic circuit. The way round this has been to change the shape of the PROM diode matrix to 256 rows by 256 columns. Thus only eight of the input lines are used to select the rows, the other five inputs select one group of eight columns from the 256. This structure is illustrated in Figure 2.13.

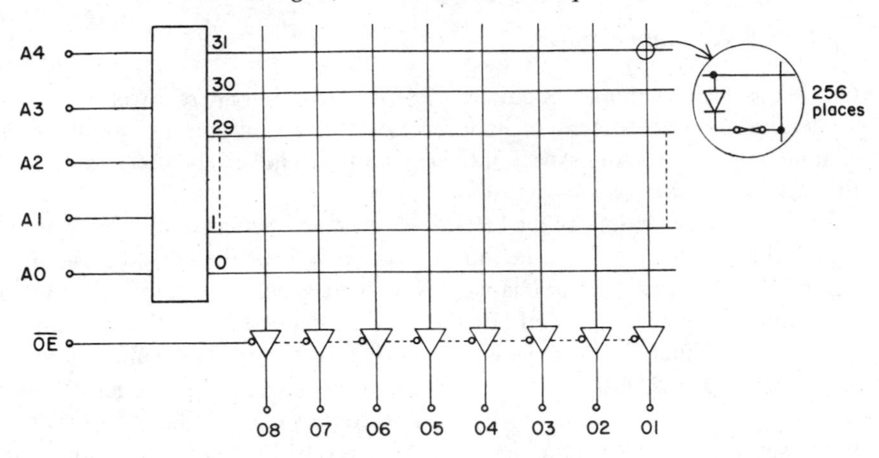

Fig. 2.12 32 × 8 PROM structure.

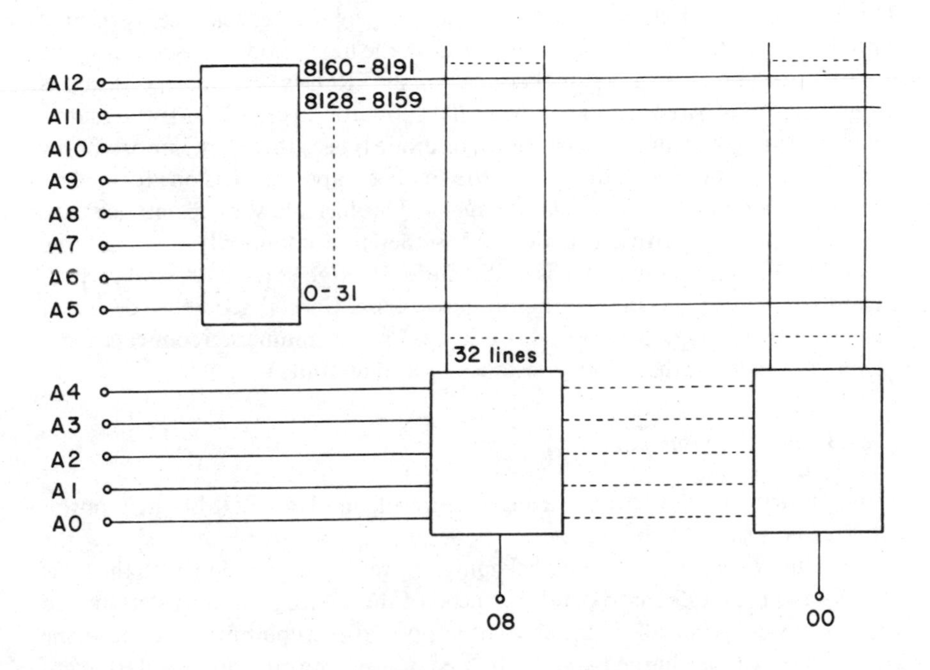

Fig. 2.13 8192 × 8 PROM fuse matrix addressing.

There is another advantage to this structure. As mentioned earlier, there is a small leakage current associated with a reverse-biased diode; if there are 8192 rows then it is possible for a column, when LOW, to have to sink the leakage current from 8191 diodes. A similar problem would exist if all the input selection were applied to the columns. Thus, there are good electrical reasons, as well as geometric ones, for making the diode array as 'square' as possible.

2.3.1.2 Speed considerations

One of the chief advantages of bipolar PROMs is that they are relatively fast, that is, it takes only a few tens of nanoseconds for the outputs to change after changing the inputs. Any aspect of the design which adversely affects the speed therefore needs careful consideration.

There are three sections of the PROM which have an effect on the delay of the signal: the first delay is from the inputs to the row lines; the second is the time taken for the column lines to change; lastly, there is delay from the columns to the outputs. Both the first and last sections are dependent on the fabrication technology in much the same way as any other logic chip. The main difference is between small PROMs which use only input row selection and larger PROMs which have extensive logic on both rows and columns. Thus a large PROM will be slower than a small PROM simply because of the fact that it contains more complex logic circuitry.

More important, in a large PROM, is the delay through the diode matrix itself. This is simply a matter of charging and discharging the capacitance associated with the diodes. Clearly, the larger the PROM, the more diodes it uses and so the matrix capacitance will be higher. In small PROMs the matrix delay is insignificant compared with the other delays, but for larger PROMs the matrix delay starts to predominate. The choice of diode type is therefore important for large PROMs. The type of diode with the smallest capacitance is one formed by a metal–silicon junction, the *Schottky diode*. This has a lower capacitance than any of the various diffused diodes and is therefore commonly used in large PROMs. Its main disadvantage is that it suffers a large voltage drop when high currents are passing, as in the programming phase. Careful geometric design of PROMs is therefore necessary and two layers of aluminium interconnection are essential to reduce other series resistance to a minimum.

2.3.1.3 Power consumption

While device speed is a very strong feature of bipolar PROMs, their power requirements cannot be ignored, the consumption being limited by the temperature rise allowable after packaging. Power is consumed in both the logic circuits and the diode matrix and the speed of the device is broadly determined by this power. After all, if the current supply to the columns is reduced the capacitance will be charged and discharged at a slower rate, and similarly with the logic circuitry.

The power can be limited by only supplying current to that part of the PROM which is actually being used. In the same way that a certain combination of input signals selects one row in the diode matrix, the input signals can be used to select which section of the PROM is being powered-up. Naturally this affects the speed to a certain extent; the signals cannot pass through unpowered sections of PROM and must wait until they are supplied, but this has less effect than reducing the current supplying each section. Figure 2.14 shows how this scheme works in practice.

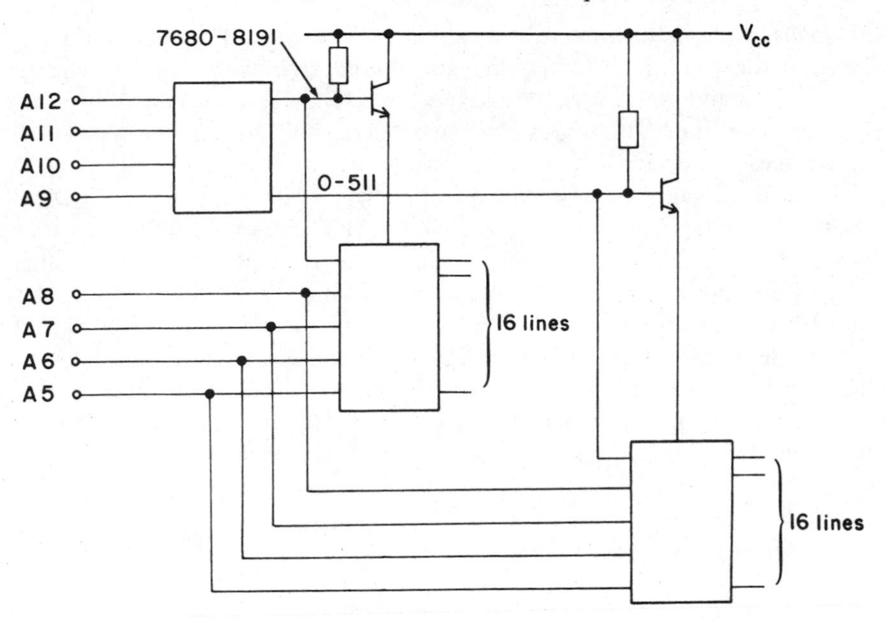

Fig. 2.14 Internal power-down in PROMs.

2.3.2 MOS PROM designs

2.3.2.1 The cell array

MOS PROMs have an even more compelling reason than bipolar PROMs for using a square memory cell array. Mention was made previously of the capacitance of the diodes, and how this became more significant with larger array sizes. The memory cell in a MOS PROM is just a small capacitor and so the array design needs particular care if performance is not to be impaired too greatly. Using a square array is one step towards minimising the capacitative loading in the memory cell array.

The usual way to reduce capacitance in integrated circuits is to make the components smaller. This has the side effect of making the signals from the memory cells smaller, so techniques more usual in RAMs (*Random Access Memories*) may be used to read the memory cells. In particular, it is possible to incorporate unprogrammed dummy cells alongside the memory array and use these as a reference voltage for a sense amplifier. The amplifier output provides the data output for the PROM.

2.3.2.2 Performance considerations

We have previously seen that speed and power are closely tied together for integrated circuit performance. Some of the techniques used in bipolar PROMs are also applicable to MOS devices. Again, only the section of PROM currently addressed needs powering, so large PROMs can work at the same current levels as smaller PROMs without a significant increase in overall power. In order to speed-up the response to address changes, internal circuitry can detect a change in the inputs and power-up all the address detectors before the new address has become established. This same signal can precharge the sensing circuits to speed-up changes in the output lines.

The main advantage that can be obtained from MOS is the use of CMOS in the peripheral logic circuits. As seen earlier, CMOS circuits consume very little power, but are still able to operate at relatively high speeds when driving other internal components in an integrated circuit. In this way, power consumption may be reduced considerably without sacrificing speed. The only adverse effect is to make the chip area larger, which increases the cost of the finished device. The larger the PROM, the smaller is the proportion of area taken by the peripheral logic, so CMOS tends to be used more for the larger MOS PROMs where the cost penalty is less.

2.3.3 Quality and reliability

2.3.3.1 Programmability

One of the parameters which cannot be tested in a fused PROM is whether or not the fuses can be programmed. The only way to test a fuse is to blow it after which it cannot, of course, be used again. The upshot is that a proportion of the PROMs will not program correctly. There are some measures which manufacturers can take to minimise this programming loss.

As far as the manufacturing itself is concerned, then, process control is the chief tool to better quality. The thickness of the fuse metal and dimensions of the fuse can be measured to ensure that the fuse can be blown by the specified current. The diffusion parameters will also ensure that the breakdown voltage is within the correct limits for the zener diodes, and for the fusing of diode fuses. Assuming that these are kept under control then correct application of the programming voltages should result in perfect programming.

In practice this will not happen because of the random manufacturing defects which can occur. These can prevent transistors from operating correctly and, for example, cause the address selection logic to operate incorrectly. Faults can also affect the metallisation causing short circuits or open circuits; these may cause the wrong fuse to be blown or a fuse not to be blown at all. These faults can be minimised by clean manufacturing conditions and by quality assurance inspections during the manufacturing process.

It should be noted that all integrated circuits are affected by these faults but it is usually possible to eliminate the rejects by electrical testing. Most PROM manufacturers now incorporate a test row and column of fuses which enable the fusability to be tested and, to a certain extent, the operation of the addressing circuits. There is also a useful spin-off from this. In an unblown PROM the outputs will remain unchanging irrespective of the inputs, so it is possible to test only one of the output levels, and impossible to test the delay time through the

PROM. If the test fuses are blown then these measurements can be made by addressing the extra memory locations.

Because the memory cells of MOS PROMs can be erased and reused, it is possible to program the device with a test pattern and then erase it before it is delivered to a customer. Naturally this will allow all the d.c. and a.c. parameters to be fully tested also. This testability means that it is not necessary to provide an extra row and column; however, there is a way in which extra memory cells can be used.

As memory arrays are made larger the chance of a manufacturing defect causing a faulty cell becomes significant. By incorporating test circuits and a register of faulty cells it becomes possible to substitute spare cells for the faulty ones during operation of the PROM. The slight increase in area is more than compensated for by the increase in the number of good PROMs which is obtained by this technique.

2.3.3.2 Data retention

Reliability of a PROM may be assessed as the length of time for which data is retained in the memory cells. All integrated circuits are subject to failure mechanisms which may stop them operating correctly at some time. Part of the process of improving integrated circuits is the reduction of the effect of failure mechanisms, thereby increasing the useful life of these devices. PROMs have additional potential failure mechanisms which must be taken care of if they are to have a useful life comparable with other integrated circuits.

Considering first bipolar PROMs, it should be noted that during normal operation there will be some current passing through the fuses. All the columns are driven by current generators and, referring back to Figure 2.6, it can be seen that this current will pass through the fuse and diode into the selected row line. If the current is significant compared with the fusing current then there will be a risk that it will blow the fuse during normal operation. Analysis of the fusing mechanism shows that there is a critical current density above which rapid melting occurs, but below which the metal melts only slowly. This well-known relationship is shown in Figure 2.15 and PROMs are designed so that the current density during operation is well below the critical level. All the major bipolar PROM manufacturers have published life test results which show that the reliability of bipolar PROMs is no different from other integrated circuit families, and that none of the failures is due to a fuse failure.

It might be thought that the shape of the fuse would have a bearing on both fusability and reliability. A bar-shaped fuse is probably easier to make because alignment is less critical during the manufacturing phase. To be effective, a notched or tapered fuse needs the narrow section to be well removed from the contact area so that fusing occurs away from the aluminium tracks. A bar fuse will always melt in the middle because heat conduction will be symmetrical and ensure that this is the hottest part. There is probably no advantage in shaping the fuse to define where the hottest part will be. Experience has shown that the process quality and electrical design will determine which PROMs are the more reliable.

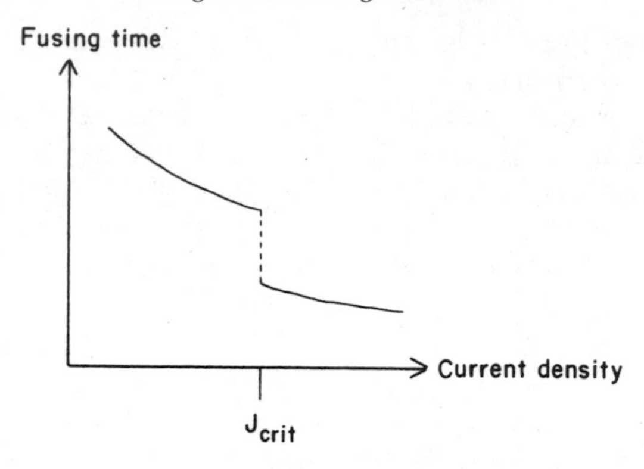

Fig. 2.15 Fusing time vs current density in nichrome.

Data integrity of MOS PROMs is dependent on different factors. As we have seen, programming is by application of a relatively high voltage to the memory cells. During normal operation the devices will not see such a high voltage so there is little risk of losing data. MOS PROMs are designed to be erased, of course, so therein lies the greatest risk of corrupting the stored data. To erase the data fully normally requires a prolonged dose of ultraviolet radiation but even a short dose will cause the threshold voltage of the memory cells to change. Erasure is a cumulative effect, that is, a continuous exposure is not required but the total accumulated time will cause the same effect.

In order to protect a MOS PROM from spurious erasure it is usual to cover the window in the package with an opaque lid or label after programming. This is because sunlight and fluorescent lights contain small amounts of ultraviolet radiation and could cause data to be lost. More important, particularly in military applications, is the effect of ionising radiations such as X-rays and even natural radioactivity. Electrons in the floating gate which are struck by a photon or particle will absorb enough energy to jump the insulating gap and start the erasing process. Even the recently introduced 'one-time programmable' MOS PROMs, which are simply EPROMs supplied in a standard opaque package, are not fully protected from these effects.

2.3.3.3 Metal fuse regrowth

A particular aspect of data retention which has caused concern to prospective PROM users in the past has been the ability of the blown fuses to heal themselves. Some very early PROMs did suffer from the problem of fuse regrowth, when a fuse which had been blown was subsequently found to be intact. It has been known for some time that metal in thin films can move under the influence of an electric field. This is one of the commonest failure mechanisms in integrated circuits and is known as electromigration.

Referring back to Figure 2.15 it can be seen that there are two regions where fuses can be blown. High current densities cause a rapid melting of the metal and surface tension pulls the metal away from the hot spot to leave a very clean gap in the fuse. Low current densities cause very local melting resulting in a ragged gap, often with islands in the middle, with small separations which can be filled again when a low voltage is applied. It is clearly important that the programming method ensures that the current density is high enough to give proper fusing. It is up to the manufacturer to design the PROM and specify the programming method, but the user must also be careful that he follows recommended procedures.

The two most important safeguards are to ensure that good contact is made to the PROM and that a PROM which does not program first time is discarded. If poor electrical contact is made to the PROM then the resulting voltage drop may be large enough to prevent the critical current density from being achieved in the fuses. Similarly, failure to program first time may be due to a faulty current generator in the programming circuit. Repeated programming attempts may cause the fuse to be blown in its unreliable mode, making it liable to regrowth during subsequent use. By following the programming specification implicitly, perfectly reliable fusing should take place.

Chapter 3

Basic Logic Structures

3.1 GATES

3.1.1 Logic operations

The basic logic operations were introduced in Chapter 1 by referring to a simple combination lock. The lock operates by setting switches to be *true* or *false*, the switches being arranged to allow a current to flow when the correct combination is achieved. A logic equation was written to describe the 'open' condition; this equation used some of the basic logic operators to connect the *variables* corresponding to the switches. In this section we shall describe the three simplest logic operators, AND, NOT and OR.

The AND operation implies that all the stated conditions must be satisfied for the result to be true. Thus, in our example, every switch must be in its correct position for the lock to open. If the combination required all the switches to be up (true) then we could write the logic equation using the symbol '*' to represent the AND operation. Sometimes the symbol '.' is used for AND but most design software packages use the '*' so we will standardise on this. The equation becomes:

$$\text{OPEN} = \text{A} * \text{B} * \text{C} * \text{D}$$

NOT is the operation which changes true to false or vice versa. It is written in logic equations by putting a line over the negated symbol, or by placing a '/' before the symbol. Thus the original equation, as defined in Chapter 1, becomes:

$$\text{OPEN} = \text{A} * \overline{\text{B}} * \overline{\text{C}} * \text{D}$$

The third of the operators is the OR function, which provides for logical alternatives. To illustrate this let us add a second combination to the lock, perhaps so that two people could have their own access to the safe! Let the second combination be A-up B-down C-up and D-down. The logic must allow either combination to OPEN the lock so we use the OR function to combine the two combinations. The symbol for OR is '+' so we can write the equation as:

$$\begin{aligned}\text{OPEN} = {} & \text{A} * \overline{\text{B}} * \overline{\text{C}} * \text{D} \\ & + \text{A} * \overline{\text{B}} * \text{C} * \overline{\text{D}}\end{aligned}$$

Figure 3.1 shows how this logic system could be built from mechanical switches. The two-way switches used in Chapter 1 are replaced by double pole two-way switches and, if either path is made, the lock will open. We will now see how

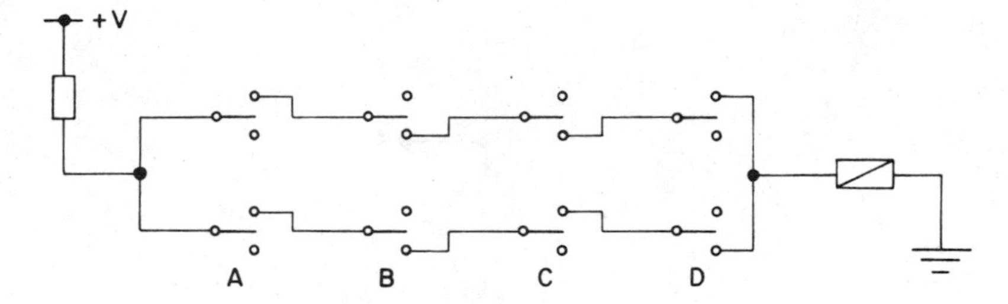

Fig. 3.1 Combination lock with switches – version 2.

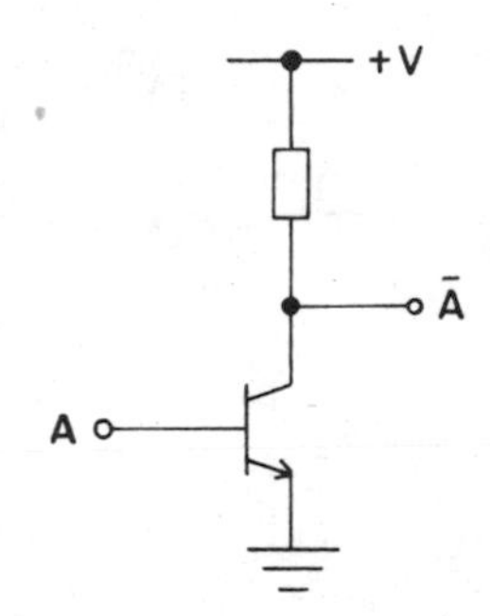

Fig. 3.2 Single transistor inverter – bipolar.

electronic switches can be used to perform the logic functions although, as implied in Chapter 1, mechanical switches are still needed to provide an interface to the outside world.

3.1.2 Physical implementation

A physical device performing a simple logic operation is called a *gate* because of the analogy between an electrical switch and a gate, which allows entry when open. Gates are formed from transistors which may be fabricated in either bipolar or MOS technology, and we will describe how both are made.

The simplest function is the NOT function, usually called an inverter. A single transistor acts as an inverter when connected to a load, as shown in Figures 3.2 and 3.3. In the case of a bipolar transistor the load is connected in the collector circuit, while the MOS transistor has it connected to the drain. For each one a HIGH voltage (i.e. close to supply voltage) will cause the transistor to conduct so that the output is a LOW voltage (i.e. close to zero volts). Conversely, a LOW input voltage turns the transistor off giving a HIGH output voltage. In CMOS the load is an active one, that is, a transistor of the opposite sense, and with both input gates connected together a LOW input voltage turns just the upper transistor on, while a HIGH voltage turns only the lower one on. The CMOS inverter is illustrated in Figure 3.4.

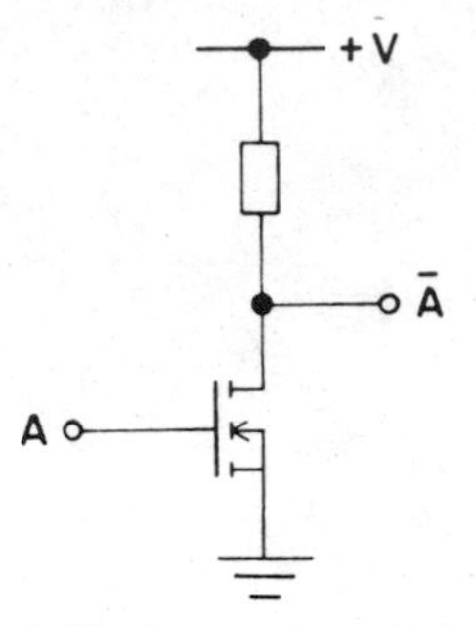

Fig. 3.3 Single transistor inverter – MOS.

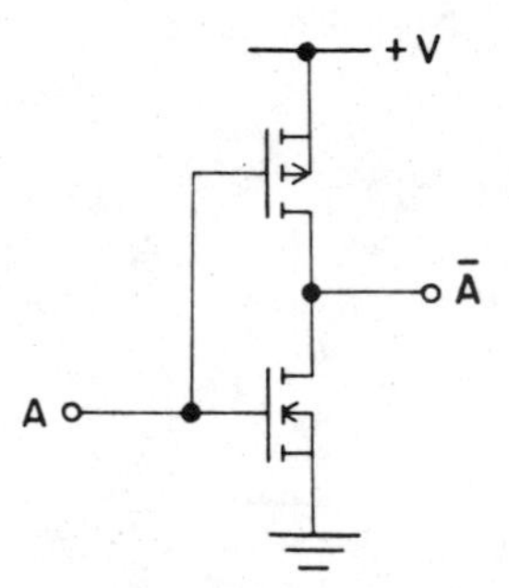

Fig. 3.4 CMOS inverter.

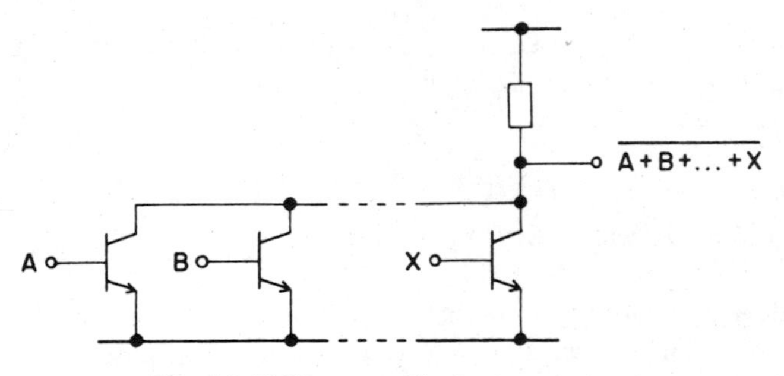

Fig. 3.5 NOR gate – bipolar transistors.

An inverter may be simply expanded to an inverting OR gate, or NOR gate, by connecting two or more transistors in parallel to the load as in Figure 3.5. Now a HIGH voltage on any of the transistor bases will cause current to flow through the load and take the output LOW. An OR gate can be made by taking the output signal through an inverter.

The situation with CMOS is not so straightforward. The lower transistors may be connected in parallel as in the bipolar circuit, but we need to examine the upper transistors, which form the load, more carefully. The purpose of the load is to pull the output towards the supply voltage when all the lower transistors are off, that is, when all the inputs are LOW. Conversely, the load needs to have a

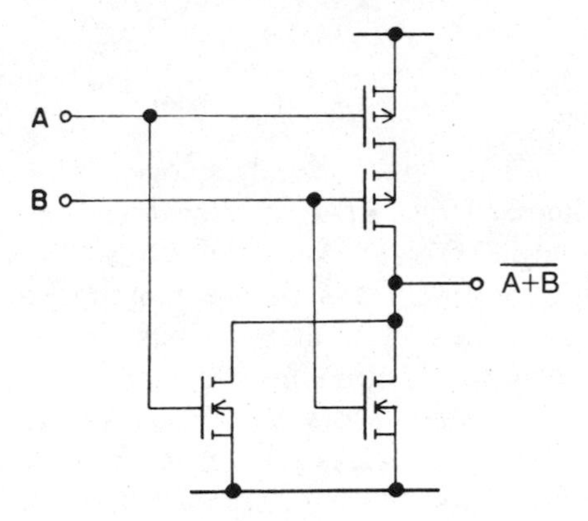

Fig. 3.6 NOR gate – CMOS.

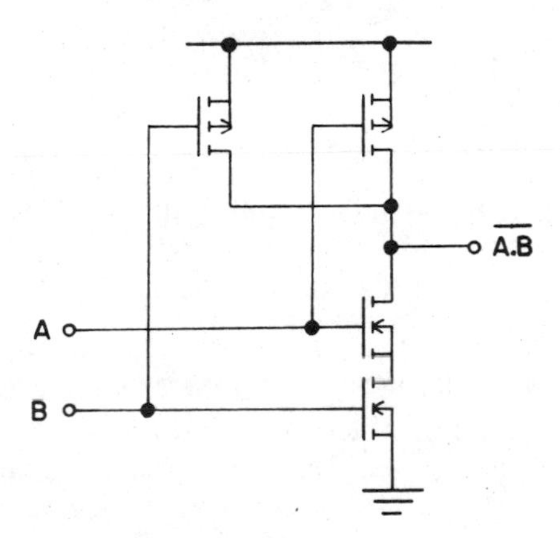

Fig. 3.7 NAND gate – CMOS.

high resistance when any of the inputs is HIGH. The configuration which meets this criterion is to connect the upper transistors in series, for then if any input is HIGH the transistor connected to it will be turned off and the load will be unable to conduct. A two-input CMOS NOR gate is shown in Figure 3.6.

To form an inverting AND gate, or NAND gate, in CMOS the NOR gate structure is simply turned upside down, as in Figure 3.7. The output must be LOW only when all inputs are HIGH, so the series structure may be used for the lower half of the gate. By connecting the upper transistors in parallel the output is pulled up to the supply voltage when any of the inputs is LOW.

Unfortunately, the series configuration may not be used in bipolar circuits, for a fairly accurate voltage of about 0.7 V needs to be maintained across the base-emitter connections to make a bipolar transistor conduct, yet not damage it with excess current. We have already seen, in Chapter 1, how diodes may be used to implement the AND function and this structure forms the basis of most bipolar logic gates.

3.1.3 Practical logic devices

3.1.3.1 'TTL' families

One of the prime considerations of a family of logic devices is that they must be capable of being connected together. That is, they all recognise the same voltages as logic HIGH and logic LOW, and when the output of one is connected to the input of another there is not a major change in output voltage because of the current flowing between the two. A number of bipolar families were introduced when integrated circuits were in their infancy but the *Transistor–Transistor–logic* or TTL family was soon established as a *de facto* standard.

Figure 3.8 shows the basic TTL NAND gate. The gating takes place in the multi-emitter input transistor, which is a compact development of the standard input diode cluster. The signal is shifted back to standard levels by the rest of the circuit, which also provides sufficient output current to drive succeeding logic stages. The standard TTL voltages are:

supply voltage – 5.0 V
HIGH voltage – >2.0 V
LOW voltage – <0.8 V

Usually there is sufficient output current to drive ten other TTL circuits, a property called *fan-out*.

3.1.3.2 'ECL' families

Although it is possible to make OR gates in the TTL families, TTL is essentially NAND-based logic. A family of circuits based on the OR structure described earlier has also been developed. This is the *Emitter-Coupled-Logic* or ECL

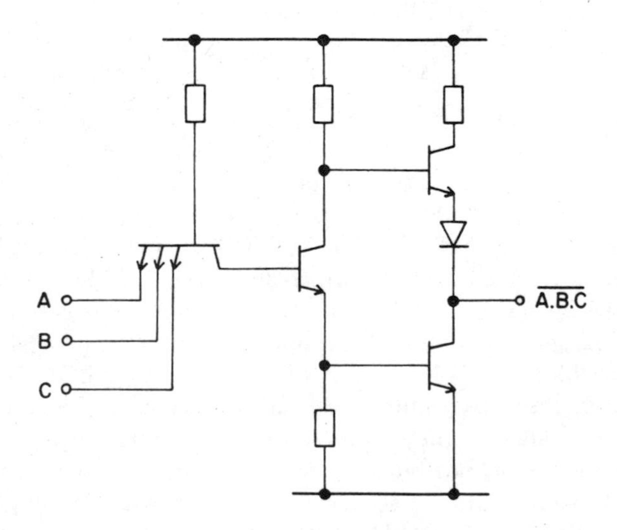

Fig. 3.8 NAND gate – TTL.

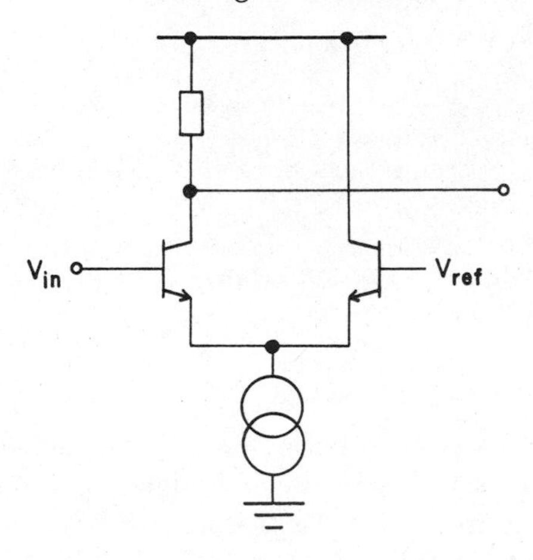

Fig. 3.9 Emitter coupled pair comparator.

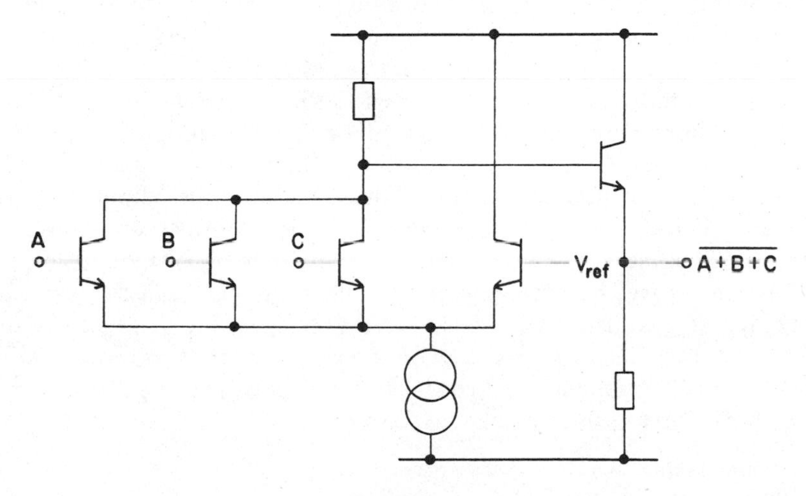

Fig. 3.10 ECL OR/NOR gate.

family. It has been refined by using a comparator principle to determine whether input voltages are HIGH or LOW. The emitter-coupled pair of transistors in Figure 3.9 form a simple comparator such that current will flow in the load if the input voltage is above the reference voltage, but not if it is below the reference. By connecting input transistors in parallel, as described above, a NOR gate can be constructed.

A complete ECL gate includes a level-shifting emitter follower, as shown in Figure 3.10, to give the correct voltage to drive the next stage. Because it uses higher current levels and a lower voltage swing, ECL gates have a shorter delay than TTL; ECL is to be found more often in higher-speed applications. Unlike most other logic families, it is powered by a negative supply, that is, the most positive voltage is 'ground', because the voltage swing across the load is more accurately defined by reference to a positive ground. The standard voltage levels are:

supply voltage – –4.5V or –5.2V
HIGH voltage – >–0.8V
LOW voltage – <–1.6V

Fan-out is quite high, but more important are the line-driving abilities of the outputs in these high-speed applications.

3.1.3.3 CMOS families

The limitations of CMOS gates are somewhat different from the bipolar logic families. The chief disadvantage is the limitation on the number of inputs to a simple gate. When turned on, a MOS transistor has a relatively high series resistance, so a load formed by transistors in series will show a high output resistance. On the other hand, the resistance of parallel transistors depends on how many of them are turned on. The latter can lead to disturbing changes in output resistance as input signals to the gate change sense. The former, however, can prevent the gate from working at all. If a CMOS NAND gate is sinking too much current, the output voltage may rise to such a level that the input voltage to the transistor on top of the string may not be enough to turn it on. Moreover, the threshold voltage in absolute terms will be different for each input.

The standard solution to these problems is to buffer the outputs with two inverters; this does not affect the logic function but does prevent conditions at the output either influencing, or being influenced by, conditions at the inputs.

CMOS does have a big advantage over bipolar circuits in that it is very tolerant to supply voltage variations. Provided that the voltage swing is sufficient to turn the transistors on and off, 'overdriving' them has no effect provided that the breakdown voltages are not exceeded. Thus, CMOS voltage levels are defined somewhat differently from bipolar, as follows:

supply voltage – 3 V to 15 V
HIGH voltage – >0.7 × supply voltage
LOW voltage – <0.3 × supply voltage

The fan-out of CMOS gates is limited more by a.c. considerations than by d.c., because the input to a CMOS gate looks like an almost pure capacitance. Driving several inputs delays signals, but makes virtually no change to their voltage level.

Although the traditional CMOS '4000' series can use the wide voltage range above, the new '74HC' family operates at TTL supply voltages and the '74HCT' version interfaces directly to TTL. Also the more complex CMOS circuits, such as PROMs and microprocessors, normally interface with TTL voltage levels.

3.1.4 Gate symbols

When drawing logic systems, standard symbols are used to represent the logic functions. There are now two systems of symbols; the traditional symbols, which

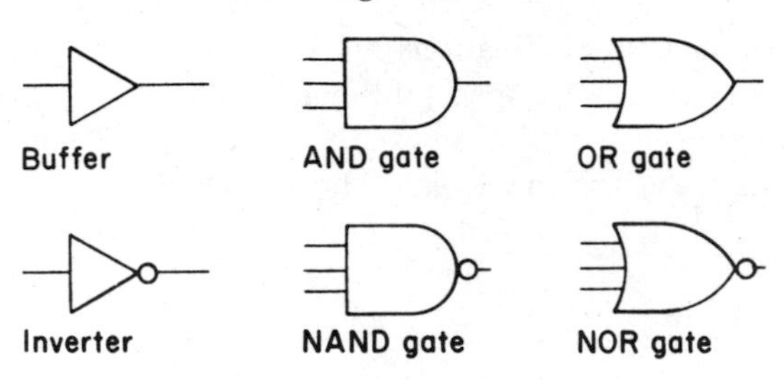

Fig. 3.11 Standard gate symbols.

we will use, and a new system devised by the IEEE/IEC. Figure 3.11 shows the symbols for the gate functions described so far. It should be noted that inversions are shown by a 'bubble' on the output of traditional symbols. An inverting bubble may also be placed on inputs when an input function is being inverted.

3.2 REPRESENTATION OF LOGIC

3.2.1 Truth tables

Until now we have represented logic functions, such as AND gates, by equations which define an output variable in terms of input variables. This may not always be the easiest way to visualise what is required from the logic function being described. For example, we wrote an equation to describe the action of a combination lock in Section 3.1.1, and we showed how this could be built from switches. Anybody wanting to operate the lock must be given the combination in some easily remembered format. The equation is not the most memorable form to retain this information. Much easier would be to describe the two combinations as 'ON, OFF, OFF, ON' and 'ON, OFF, ON, OFF', both of which have some pattern to them. This is the basis of the truth table.

As its name implies, it is a table of true conditions. In electronic logic systems we usually refer to 'HIGHs' and 'LOWs', rather than ON and OFF, and use the abbreviations 'H' and 'L' to refer to them in truth tables. The truth table representation of the combination lock is shown below:

A	*B*	*C*	*D*	*OPEN*
	Active Level			*H*
H	L	L	H	A
H	L	H	L	A

There are, of course, sixteen possible combinations of H and L for the four inputs but only the two true, or active, combinations have been listed, it being understood that the others are inactive. The active output level has been defined as 'H' so it is implied that all other combinations will give a LOW output.

In some cases all the combinations are defined for the sake of clarity or completeness but then, to save space, the symbol 'X' or '–' is used to represent an input where either 'H' or 'L' will have the same effect. An example of this is the complete truth table for an OR-gate, shown below:

A	*B*	*C*	*Y*
H	–	–	H
–	H	–	H
–	–	H	H
L	L	L	L

It is implicit in the OR function that if input A is HIGH then the output will be HIGH no matter what the condition of the other inputs; but the output will only be LOW if all inputs are LOW.

3.2.2 Karnaugh maps

Another way of representing a logic function, and providing a very useful tool for manipulating logic, is the Karnaugh map. A blank Karnaugh map for four input variables is shown in Figure 3.12. First of all it should be noted that the map contains sixteen squares, that is, one for each possible combination of the input variables. Secondly, it should be seen that the four variables are split into two groups of two, and each row or column of the map corresponds to a unique combination of the two variables. Thus, the combination corresponding to each square is defined by the combination of the variables in the row and column meeting in that square. The square containing the asterisk is defined by A–'H' B–'L' C–'H' D–'L'.

What is not so obvious is the way in which the rows and columns are arranged. The sequence of the combinations of variables is 'LL' 'LH' 'HH' 'HL', which is a method of counting known as *Gray code*. The key to counting in Gray code is that only one variable changes at any one time, and no combination is repeated. A Karnaugh map can be set up, in principle, for any number of input variables by writing the Gray code along the row and column axes and entering the output condition for each combination of input variables in the appropriate square in the map. Naturally, this will be rather tedious for more than eight or ten variables although there are computer programs capable of handling larger maps.

3.2.2.2 Using Karnaugh maps

Having seen how to create a Karnaugh map for a logic function we will now look at some of the ways in which they can be used. Figure 3.13 shows the Karnaugh map for our combination lock. The two squares with 'H's represent the two AND gates in the circuit diagram; in fact, each square can be physically implemented by means of an AND or NAND gate, depending on whether it contains an 'H' or 'L'. Thus, it shows immediately how a function may be

D	C	A=0, B=0	A=0, B=1	A=1, B=1	A=1, B=0
0	0				
0	1				*
1	1				
1	0				

Fig. 3.12 Karnaugh map for 4 variables.

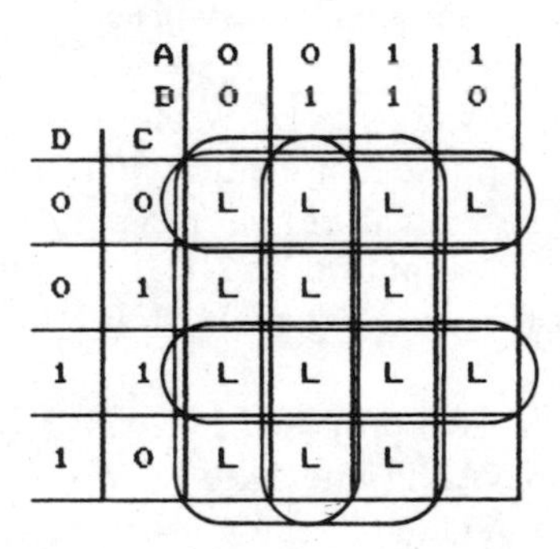

D	C	A=0, B=0	A=0, B=1	A=1, B=1	A=1, B=0
0	0				
0	1				H
1	1				
1	0				H

Fig. 3.13 Karnaugh map – combination lock.

D	C	A=0, B=0	A=0, B=1	A=1, B=1	A=1, B=0
0	0	L	L	L	L
0	1	L	L	L	
1	1	L	L	L	L
1	0	L	L	L	

Fig. 3.14 Karnaugh map – combination lock ('L's grouped).

constructed from simple gates. It may also be used to improve the performance of a design.

In one of the fastest TTL families, 74FAST, NAND gates have a maximum delay time of about 5.5 ns, while the figure for AND gates is about 6.5 ns. Thus, the combination lock could be made about 1 ns faster by using fourteen NAND gates driving an AND gate instead of two AND gates driving an OR gate. The AND and OR gates have the same speed but AND is needed instead of OR because we are looking for an absence of LOWs rather than a HIGH present. This might not be very efficient in terms of the number of circuits used by the design but it might be a crucial factor in some designs which are critical on speed. The Karnaugh map may help even further.

If we look at the top right-hand corner of Figure 3.14 we can see that there are two 'L's in adjacent squares. These correspond to the functions $A*B*\overline{C}*\overline{D}$ and $A*\overline{B}*\overline{C}*\overline{D}$, which implies that the variable B can be either HIGH or LOW without affecting the output. These two functions can therefore be replaced by the single function $A*\overline{C}*\overline{D}$. This is an example of a general principle which can be applied to Karnaugh maps called *logic minimisation*. Any pair of adjacent squares containing identical outputs may be combined into a single function, and this principle may be extended to quartets, octets and so on indefinitely. Figure 3.14 shows the Karnaugh map with the fourteen 'L' functions combined into two octets and two quartets, that is, four gates in all, which could make it worth while to achieve the improved performance.

A further point about logic minimisation can be illustrated here. The four gates could equally well be achieved by combining the two left-hand columns into an octet, the third column as a quartet and leaving the two 'L's in the right-hand column as single functions. There are two reasons for proceeding as we did. Firstly, the two octets are physically implemented as inverters and the two quartets as two-input gates, so the logic devices required are much simpler, leading to a more compact solution. Secondly, the alternative solution could give rise to spurious signals, known as *glitches*, when inputs change without causing an intentional change in the outputs.

Consider the two input combinations $A*B$ and $A*\overline{B}*\overline{C}*\overline{D}$ with A HIGH, C and D LOW, and B changing from LOW to HIGH. To start with only $A*\overline{B}*\overline{C}*\overline{D}$ is true so the output AND gate will see HIGHs from the other three NAND gates and a LOW from $A*\overline{B}*\overline{C}*\overline{D}$, so it will have a LOW output. At the finish only $A*B$ is true so it will put a LOW on to the AND gate and its output will remain LOW. Now suppose that the delay through $A*B$ is longer than the delay through $A*B*\overline{C}*\overline{D}$; there will be a short period when all four NAND gates have a HIGH output so the output will go HIGH momentarily. This short glitch could be interpreted as a valid signal by following circuitry and cause a malfunction.

The solution which we have proposed overcomes this problem by ensuring that there are at least two NAND gates with LOW outputs either before or after a change in a single input. A glitch-free solution may always be ensured if groups of variables are overlapped wherever possible, as in this case.

3.2.3 Boolean algebra

We have already seen that logic systems may be represented by equations, as in our combination lock. Just as there is a set of algebraic rules for manipulating mathematical equations, there is also a set of rules for working with logic equations. They were first postulated by Boole, so the process of manipulating logic equations is commonly referred to as *Boolean algebra*.

In many ways Boole's rules and theorems mirror those of mathematics. For example, the order in which variables are written is not significant for:

$A*B = B*A$
$A+B = B+A$

correspond exactly to their algebraic equivalents. The rules for expanding brackets, however, are not quite as straightforward:

$A*(B+C) = A*B + A*C$ looks obvious enough but
$(A+B)*(A+C) = A + B*C$ needs some justification

By drawing-up a truth table to derive the left-hand side and right-hand side of the equation, we can show that they are identical:

A	*B*	*C*	*A+B*	*A+C*	*LHS*	*B*C*	*RHS*
0	0	0	0	0	0	0	0
0	0	1	0	1	0	0	0
0	1	0	1	0	0	0	0
0	1	1	1	1	1	1	1
1	0	0	1	1	1	0	1
1	0	1	1	1	1	0	1
1	1	0	1	1	1	0	1
1	1	1	1	1	1	1	1

The most powerful rules in Boolean algebra are known as DeMorgan's Laws and deal with the relationship between the AND, OR and NOT functions. They are as follows:

$$(\overline{A+B+C+D+}\ldots) = \overline{A}*\overline{B}*\overline{C}*\overline{D}* \ldots$$
$$(\overline{A*B*C*D*}\ \ldots) = \overline{A}*\overline{B}*\overline{C}*\overline{D}* \ldots$$

Once again a truth table may be used to justify the laws, but this will be left as an exercise for the reader. However, we can show these rules being applied to our combination lock equations to perform the same transformation that we achieved in the previous section with the Karnaugh map. We start with the equation:

$$\begin{aligned}
OPEN &= (A*\overline{B}*\overline{C}*D) + (A*\overline{B}*C*\overline{D}) && \\
&= (\overline{\overline{A}+B+C+\overline{D}}) + (\overline{\overline{A}+B+\overline{C}+D}) && \text{– DeMorgan's first law} \\
&= ((\overline{\overline{A}+B+C+\overline{D}}) * (\overline{\overline{A}+B+\overline{C}+D})) && \text{– DeMorgan's second law} \\
&= (\overline{\overline{A} + B + (C+\overline{D})*(\overline{C}+D)}) && \text{– expanding brackets} \\
&= (\overline{\overline{A} + B + C*(\overline{C}+D) + D*(\overline{C}+D)}) && \text{– expanding brackets} \\
&= (\overline{A + B + C*D + \overline{C}*\overline{D}}) && \text{– expanding brackets}
\end{aligned}$$

These are just the four terms which we created by grouping '0's in the Karnaugh map of the same function.

3.3 COMPLEX LOGIC FUNCTIONS

3.3.1 Combinational functions

3.3.1.1 Multiplexers and decoders

When discussing ROMs we described the need for a circuit which would select a particular combination of HIGHs and LOWs from the address inputs, and to apply a signal to the appropriate row or *word line* of the ROM. The circuit function which does this is called a *decoder*. We also established the need to select one of a number of columns or *bit lines* to drive the outputs, again according to the combination of address input signals. A *multiplexer* is the function required to do this. Now that we have examined the various ways of representing logic functions we can see how these and other functions may be synthesised.

To start with we shall look at *combinational* functions, in which input signals are combined according to a logic requirement to give a predefined output signal. It does not matter in which order the signals are applied, or what the output state was before the inputs were changed; a certain combination of inputs will always lead to the same output condition. Multiplexers and decoders are examples of combinational logic functions.

We have already seen an example of a decoder in our combination lock circuit, where a given combination of input signals will cause an AND gate to go HIGH, if the LOW signals are inverted first. A decoder, then, needs an AND gate for each output, with every input signal, or its complement, being gated together. If there are n input signals then there are 2^n possible combinations of inputs. There must be 2^n AND gates in an n-input decoder. This is illustrated in Figure 3.15 which shows a three-input, or eight-bit, decoder. The inputs are *buffered* so that each presents only a unit load to the driving circuits. There may be a fourth input which can be used as an *enable* line; by connecting it to every gate in the circuit it must be HIGH for the circuit to operate at all. This makes it possible to combine several small decoders into a large decoder.

The multiplexer in Figure 3.16 uses exactly the same principle as the decoder. The three address lines select one of the AND gates because only one combination of HIGHs and LOWs is true for each gate. A separate data input is the fourth input to each gate so the output of the selected gate reflects its data input. An enable line could be a fifth input to each gate. All the deselected gates have a LOW output so only the output from the selected gate can affect the output from the OR gate. The multiplexer output is thus the same as the selected input and is not affected by non-selected inputs.

3.3.1.2 The exclusive-OR function

The exclusive-OR function forms the basis of many arithmetic circuits and so merits a separate mention. Exclusive-OR could be called the inequality function for it is defined as being true when its two inputs are different from each other. In other words, its truth table is:

A	B	A :+: B
L	L	L
L	H	H
H	L	H
H	H	L

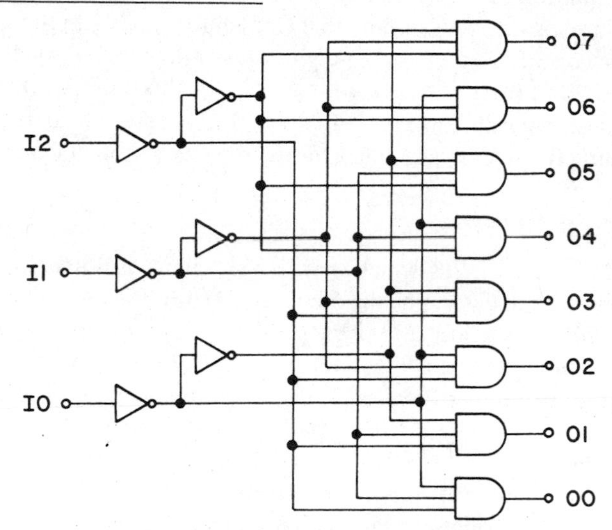

Fig. 3.15 8-bit decoder circuit diagram.

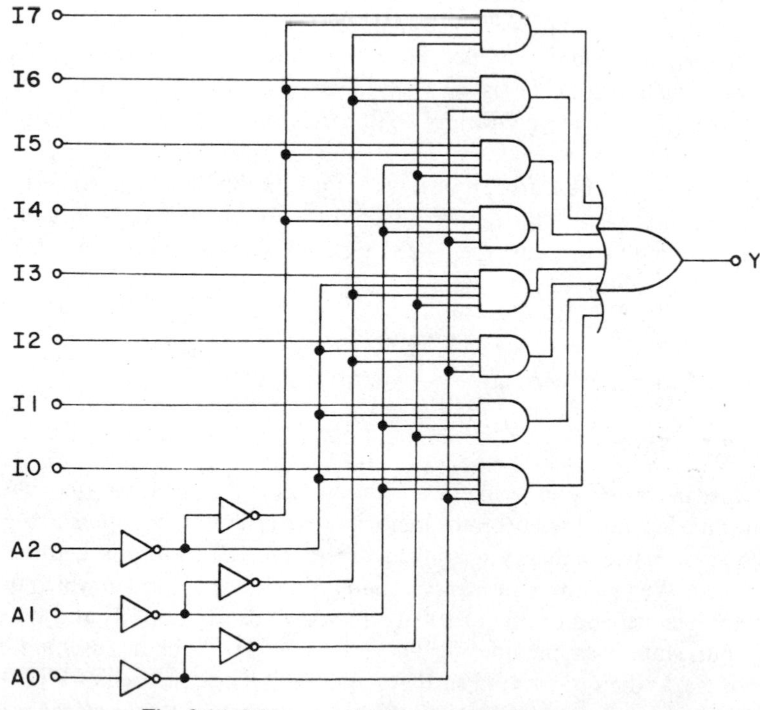

Fig. 3.16 8-input multiplexer circuit diagram.

The symbol ':+:' is one way of representing exclusive-OR in equations. If we look at the Karnaugh map for exclusive-OR in Figure 3.17 we can see that it may be built from AND and OR gates according to the equation:

$$A :+: B = \overline{A}*B + A*\overline{B}$$

Further examination of its truth table shows why it is important in arithmetic rather than pure logic circuits, although it frequently appears in those as well. A and B may be considered to be single-digit binary numbers by calling the LOW state binary '0' and the HIGH state binary '1'. In this case A:+:B is the least significant bit of the arithmetic sum of A and B. Exclusive-OR gates are widely used in building full adders and other more complex arithmetic circuits.

3.3.1.3 Parity generators

One use of the exclusive-OR which is worth dwelling on further is the parity generator. It is used to determine the number of HIGHs in a logic function, and an exclusive-OR gate is just a two bit parity generator. An even number of HIGHs in the function yields a HIGH output while an odd number yields a LOW output. Parity is used as a simple check on the integrity of data by seeing if the measured parity agrees with a parity bit associated with the data and calculated earlier.

We may try to build a parity generator from AND and OR gates by mapping the function on to a Karnaugh map, with a typical result as in Figure 3.18. The resulting 'checkerboard' pattern is the same however many inputs are considered; for the 6-input case illustrated, thirty-two AND-gates and a thirty-two-input OR-gate would be needed. However, a much neater solution is obtained by using exclusive-OR gates as shown in Figure 3.19. The function can thus be built from just five exclusive-ORs and an inverter, less than half the hardware of our first attempt.

The lesson to be learnt from this example is that a single-minded approach does not always give the best result. In this case the high incidence of diagonal outputs would hint to an experienced logic designer that exclusive-OR might lead to a better solution than simple gates.

3.3.2 Sequential functions

3.3.2.1 Storage elements

Earlier in this chapter we defined a combinational function as one in which the outputs depended solely on the logic signals present at the inputs irrespective of the order in which they were applied, or the state of the outputs before they were applied. We can now consider a second type of logic device in which the output state does depend on the order in which signals are applied and on what the output state was previously. These devices are called *sequential functions* because, as their name implies, they depend on the sequence in which things are done to them. In order to achieve this they must have *storage elements* built into

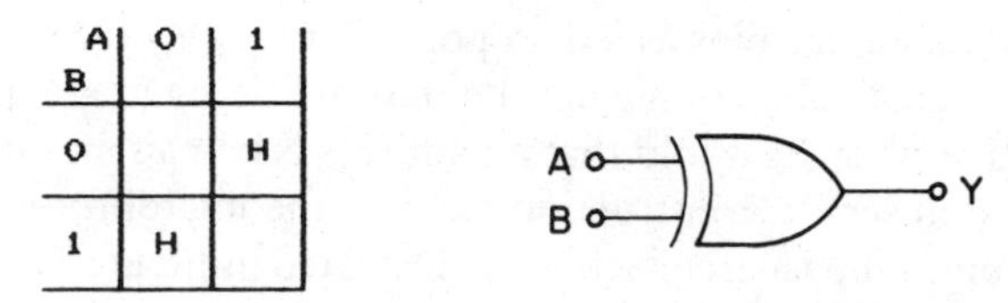

Fig. 3.17 Karnaugh map – 2-input exclusive-OR gate.

		A	0	0	0	0	1	1	1	1
		B	0	0	1	1	1	1	0	0
		C	0	1	1	0	0	1	1	0
D	E	F								
0	0	0	H		H		H		H	
0	0	1		H		H		H		H
0	1	1	H		H		H		H	
0	1	0		H		H		H		H
1	1	0	H		H		H		H	
1	1	1		H		H		H		H
1	0	1	H		H		H		H	
1	0	0		H		H		H		H

Fig. 3.18 Karnaugh map – 6-input parity generator.

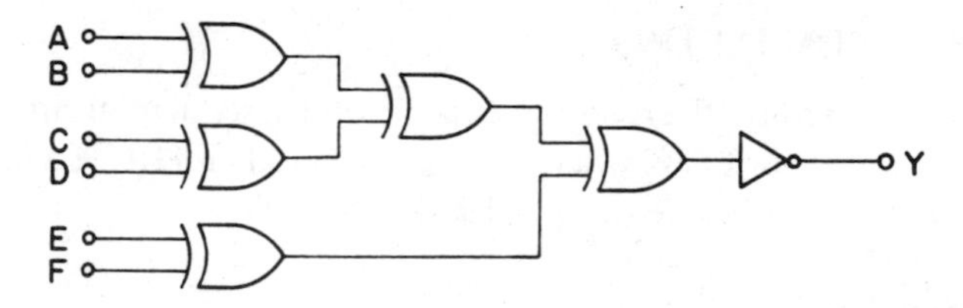

Fig. 3.19 6-input parity generator circuit diagram.

them so that they can 'remember' their immediate past history.

To illustrate this we can look again at a typical microprocessor architecture. In Chapter 2 we described a particular architecture in which the data, address and instruction busses were completely separate. This type of structure leads to integrated circuits with a very high pin count so it is much more common for microprocessors to share pins for some functions; very often it is the data bus which is shared with some of the address lines. For part of the time the microprocessor is sending out the address of the next instruction while for the remainder it is using the pins as a data port. Fortunately for the user there will be a separate signal telling him what the pins are doing at any particular time. This means that when he is told that an address is being sent out he must store it, so that he can send the instruction back to the microprocessor at the right time. This more compact architecture is illustrated in Figure 3.20, together with a timing diagram showing the systems requirements.

In this example we have assumed that the processor works in four 'phases': in phase 1 it requires an instruction, in phase 2 it carries it out internally, in phase 3 it outputs the new address for the next instruction, while phase 4 is used for data transfer to other parts of the system. During phase 3 a signal called ALE (address latch enable) goes HIGH to indicate that that is the time when the address is being given out.

3.3.2.2 Basic D-latch

We will now see how to build a circuit element capable of storing one bit of the address. Its output must be HIGH when ALE is HIGH and when the input is HIGH; it must also be HIGH when ALE goes LOW and the output is already HIGH. This is achieved by feeding the output back as another input and gating it with $\overline{\text{ALE}}$. The logic equation is thus:

$$\text{OUT} = \text{IN}*\text{ALE} + \text{OUT}*\overline{\text{ALE}}$$

The output can change only when ALE is HIGH because when ALE is LOW the output is feeding itself; it is said to be *latched*. The usual symbols for input and output are 'D' (for Data) and 'Q' and this type of latch is called a *D-Latch*. If we look at the Karnaugh map for the D-latch in Figure 3.21, we will see that the design is not quite complete. The two AND terms in the logic equation do not overlap so there is a risk of causing a glitch. In order to guard against this we must include a third term giving the complete equation as:

$$\text{Q} = \text{D}*\text{LE} + \text{Q}*\overline{\text{LE}} + \text{D}*\text{Q}$$

Note that this circuit fulfils the requirements of being a sequential circuit because Q can be either HIGH or LOW when LE is LOW and D is HIGH depending on whether D went HIGH before or after LE went LOW.

3.3.2.3 D-type flip-flop

One of the problems with combinational logic is the propagation delay through the logic elements, and we have already seen how this can lead to glitches if proper care is not taken with the design. One way to get round this problem is to sample the logic functions only at fixed intervals so that data moves through the system in discrete jumps, and it does not matter what happens between those jumps. The D-latch is not suitable for this purpose because data can move from input to output at any time when LE is HIGH. By putting two D-latches in series we can create the function which we require, as shown in Figure 3.22. If the first latch has an inverter in its enable input (i.e. $\overline{\text{LE}}$) then data can reach Q1 when LE is LOW. Once LE goes HIGH the data is passed to the output of the second

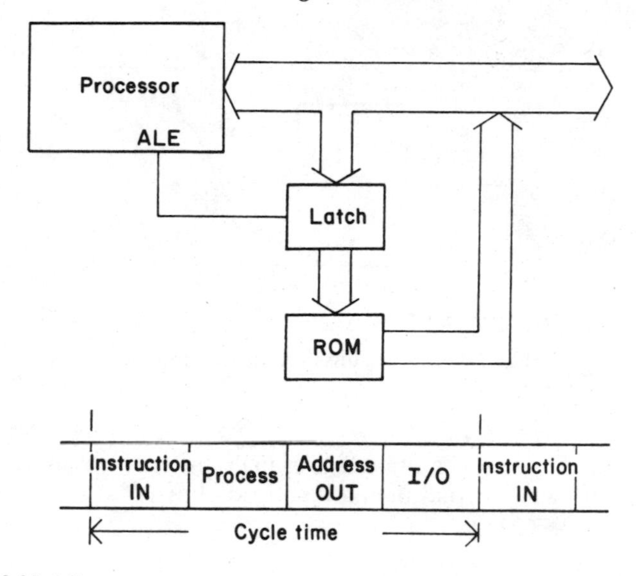

Fig. 3.20 Microprocessor architecture using multiplexed address/data bus.

D / Q / LE	D=0, Q=0	D=0, Q=1	D=1, Q=1	D=1, Q=0
0		H	H	
1			H	H

Fig. 3.21 Karnaugh map – D-latch.

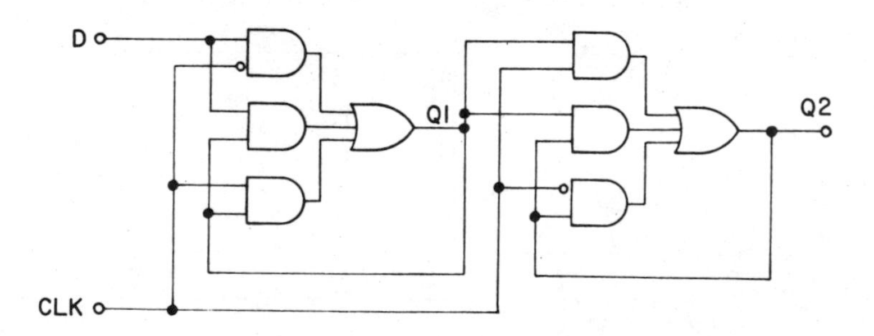

Fig. 3.22 D-type flip-flop circuit diagram.

latch, Q2, but Q1 is now fixed until LE goes LOW again. Thus data is transmitted through the two latches at the moment when the enable changes from LOW to HIGH.

The first latch is usually called the master and the second is called the slave, the whole circuit being a *master–slave D-type flip-flop*. Because the most common

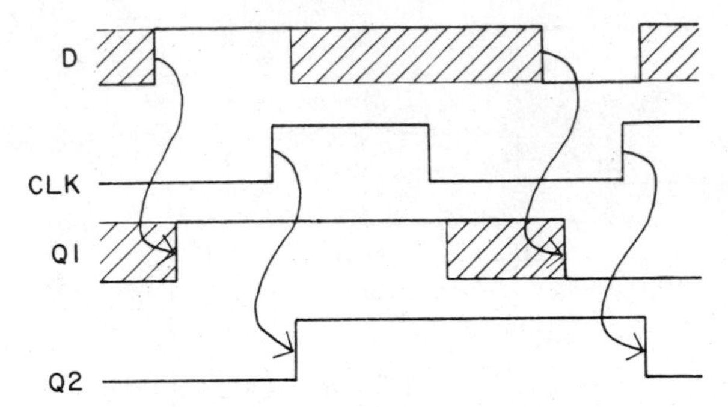

Fig. 3.23 D-type flip-flop internal timing.

use of the flip-flop is to synchronise logic signals, the enable is more commonly called the clock. The internal timing described above is shown in the timing diagram in Figure 3.23.

3.3.2.4 Summary of flip-flop types

The D-type flip-flop described above is only one of many types, all with different properties and additional features. The designs for some of these are given in Chapter 8 but a summary of their features will be given here in the form of truth tables so that we may refer to them further on in the book. In the truth tables the use of lower-case letters refers to the state of the signal in question before the clock or enable signal is applied.

(i) *D-latch*	*D*	*LE*	*Q*
	H	H	H
	h	L	H
	L	H	L
	l	L	L

(ii) *D flip-flop*	*D*	*Clk*	*Q*
	h	^	H
	l	^	L

(iii) *R–S latch*	*R*	*S*	*Q*
	L	L	q
	H	L	L
	L	H	H
	H	H	undefined

(iv) *R–S flip-flop*	*R*	*S*	*Clk*	*Q*
	L	L	^	q
	H	L	^	L
	L	H	^	H
	H	H	^	undefined

(v) *J–K flip-flop*	*J*	*K*	*Clk*	*Q*
	L	L	^	q
	H	L	^	H
	L	H	^	L
	H	H	^	$\overline{q}$

It is also quite common for an R–S latch to be combined with flip-flops to provide an asynchronous 'setting' and 'resetting' feature.

3.3.3 Timing considerations

3.3.3.1 Combinational functions

We have already noted that there is a delay between data being applied to the inputs of a logic circuit and the resulting output change. This is due to circuit capacitances being charged and discharged, so it follows that identical circuits made on a common process will have the same delay time. In practice, process variations make this only approximately true but it is common to quote a figure for the *gate delay* for most logic families. We have already seen how glitches may occur by different logic paths having different delays before being combined in a single output. This effect can sometimes be eliminated by careful design, such as overlapping groupings in a Karnaugh map; nevertheless, it is good practice to try to ensure that a minimum discrepancy in number of gate delays occurs between different paths in a logic circuit, that is, that all signal paths contain the same number of gates.

Sometimes it is necessary to ensure that one signal path is longer than another so that signals arrive in the correct sequence. This is often achieved by inserting extra gates or inverters into the 'slower' path. Correct operation may often ensue from this technique but there is always a danger that things can go wrong. Although maximum delays are always specified in logic device data, minimum delays frequently are not. Further, the actual delay will also depend on factors such as stray capacitance, supply voltage and temperature, and a positive difference under one set of circumstances can become a negative difference under a changed set. The only safe way to proceed in these circumstances is to use sequential logic.

3.3.3.2 Latches

If we look back to the equation for a D-latch we can see that the two input signals, 'data' and 'enable', both suffer two gate delays between input and output. The circuit diagram, Figure 3.24, shows this more graphically and also shows that LE has two different paths, one inverted and the other non-inverted. If D changes while LE is HIGH then the output will change after a predictable delay. If LE goes LOW while D is unchanging then the output will be latched into a predictable state. Consider what happens, though, when LE goes LOW while D is also changing; the state of the output is now not easily predicted.

To be sure about the final outcome, D must be stable around the time when LE is changing from HIGH to LOW. Part of the specification of a latch must include a definition of the time for which D must not change, but we can estimate what this must be by analysing the circuit. If we start with D and LE both HIGH, then AND gates 1 and 2 are both HIGH and gate 3 is LOW. In the latched condition, when LE is LOW and D can be LOW, only gate 3 need be HIGH. Thus to ensure reliable operation of the latch we must be certain that gate 3 goes HIGH before gates 1 and 2 can go LOW. Taking LE LOW will send gate 1 LOW so Q must be HIGH before LE goes LOW. D must therefore be HIGH for at least two gate delays before LE goes LOW. This is called the *set-up time*. Because LE is inverted before being gated with Q there is an extra gate in this path, so D must stay HIGH for one gate delay after LE goes LOW. This is called the *hold time*. These timings are illustrated in Figure 3.25.

The third parameter which must be defined is the length of time for which LE

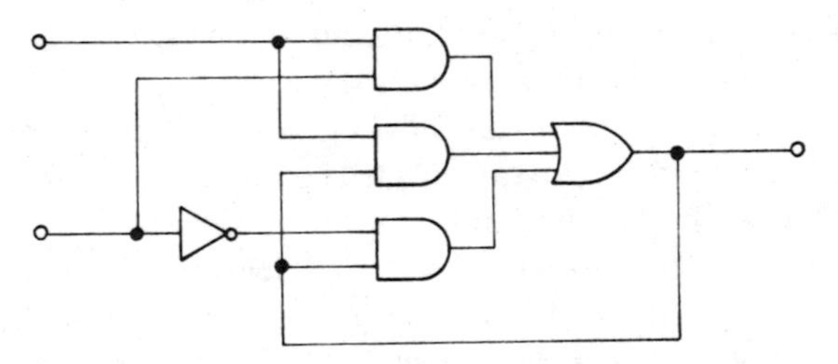

Fig. 3.24 D-latch circuit diagram.

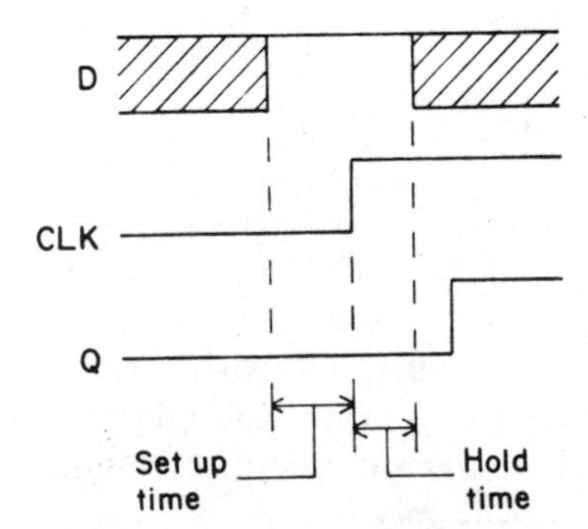

Fig. 3.25 Flip-flop timing definitions.

must be HIGH to allow D to be reliably latched. If LE and Q are LOW then all three AND gates are LOW irrespective of D. In order to latch a HIGH, LE must stay HIGH for long enough for a HIGH Q to be fed back before LE goes LOW, otherwise there is a chance that all three gates will go LOW again. This minimum pulse width is clearly equal to two gate delays to allow the feedback to become established.

3.3.3.3 Flip-flops

A simple master–slave flip-flop is merely two latches in series, as we have already established. Reliable operation depends on data being properly latched into the master section of the flip-flop, therefore the same criteria apply to the master as to the simple latch. These same criteria must also be extended to the slave section for the master Q becomes the slave D. Master Q must therefore be established two gate delays before the active clock edge so the set-up time becomes four gate delays. The hold time is not affected because the delay through the master ensures that Q will remain stable long after the active clock edge. Because the master and slave use opposite senses of the clock signal for latching, both HIGH and LOW clock states are subject to minimum widths.

The set-up time of four gate delays is rather long to make an efficient practical flip-flop, so these use R–S latches for the most part. The R–S latch has only a single gate delay between input and output so the set-up time for a flip-flop is reduced to two gate delays. A further advantage is that it is relatively simple to build-in asynchronous set and reset by adding a parallel latch so a more versatile, as well as a faster device is obtained.

3.3.3.4 Metastability

Having established rules for the timing of input signals to latches, we can now see what happens if those rules are broken. In Figure 3.26 we have assumed that the D-input to a D-latch goes HIGH one gate delay before LE goes LOW. The effect of this is to send gate 1 HIGH for one gate delay. Q therefore also goes HIGH but for just one gate delay and this pulse is fed back to gates 2 and 3 sending them HIGH, again for one delay period only. This alternating HIGH and LOW would carry on indefinitely if the gate delays were all exactly equal. In practice

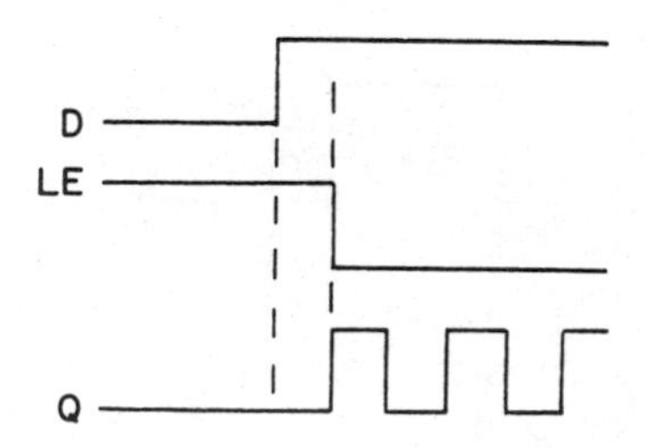

Fig. 3.26 D-latch timing violation.

they are not, but the closer they are the longer it will take for the oscillation to die away.

In a flip-flop the effect of the master oscillating, when it should be latched, will be passed directly through the slave whose latch is open. Stray capacitance may well cause the oscillation to be smoothed at the device output so the apparent effect will be for the output to lie between the HIGH and LOW states for a relatively long time before a normal level is restored. In extreme cases, where the mark–space ratio of the oscillation matches the difference in gate delays internally, this *metastable state* can continue almost indefinitely.

The lesson is that set-up and hold times of flip-flops should not be violated as it can lead to unpredictable results. Unfortunately this is not always possible, for one of the most common uses of flip-flops is to synchronise asynchronous signals. In this case, the possibility of metastability should be recognised and the system constructed so that there is an overwhelming probability that the flip-flop will have left its metastable state before the signal needs to be used.

3.4 SEQUENTIAL LOGIC SYSTEMS

3.4.1 System examples

3.4.1.1 Registers

While a latch or flip-flop is capable of storing a single bit of data it is more usual to want to store several bits. For example, in the architecture of Figure 3.20 the address of the next instruction has to be held until the microprocessor is ready for it. A group of flip-flops all controlled by the same clock signal is called a *register*. In this case we have a set of latches rather than flip-flops but the same principle applies. Because each latch has its own input and output it is an example of a parallel register. Parallel registers have some disadvantages, particularly if being made as a single integrated circuit. The most obvious is the number of connections required, each connection having to be made via a separate pin. For a register of n flip-flops there will be $2n+1$ connections.

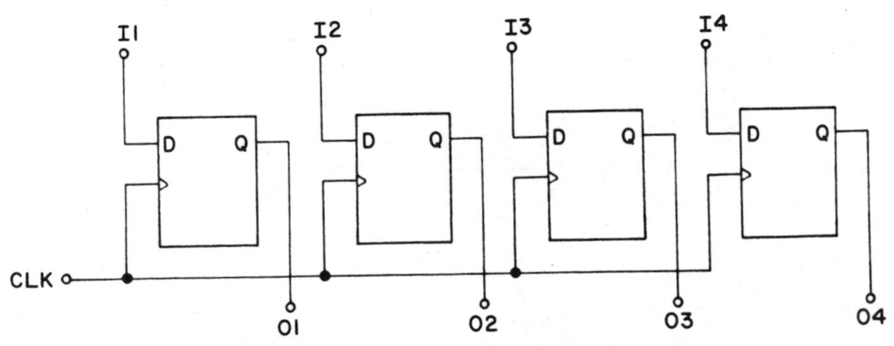

Fig. 3.27 4-bit parallel register block diagram.

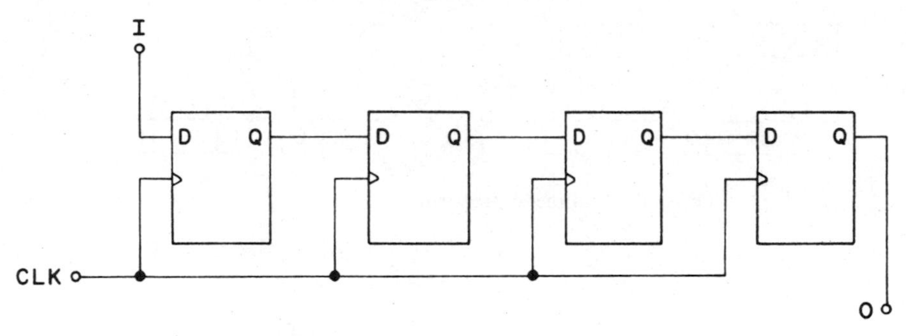

Fig. 3.28 4-bit shift register block diagram.

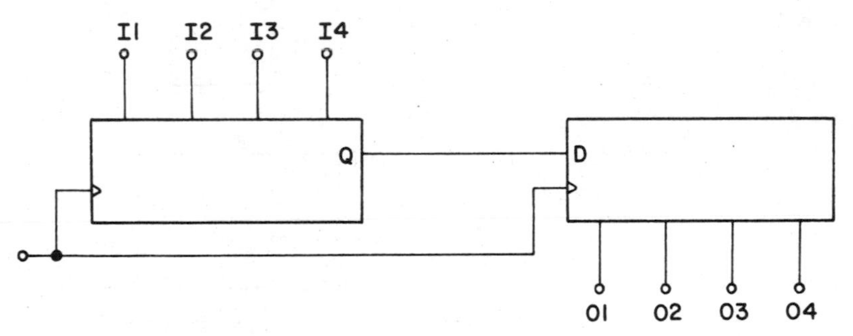

Fig. 3.29 Single wire communication between shift registers.

Furthermore, to connect the register to another device will require one wire for each register bit.

The block diagram of a four-bit parallel register is shown in Figure 3.27. Suppose that the output of each flip-flop is fed to the input of the flip-flop on its right-hand side, we then have the arrangement of Figure 3.28. On a clock pulse the input to the left-hand flip-flop is loaded, but the previous contents are loaded into the next one, and so on down the line. The data is thus shifted along the register, hence the name *shift register* or *serial register* for this arrangement.

The number of connections is now reduced to three, irrespective of the number of register bits. The chief disadvantage is that *n* clock pulses are needed to load the register or to read its contents. It is thus much slower to use than a parallel register. Two shift registers can transfer their data along a single connecting wire so hybrid versions may be used to save wiring when two parallel systems need to communicate. In Figure 3.29 the left-hand register is loaded as a parallel register and the data output as a shift register; the right-hand register is loaded with serial data and read out in parallel.

Another application of shift registers is in arithmetic circuits. If the bits loaded into the register represent a binary number, then shifting them one place results in the number divided by two. Similarly, if the connections are reversed so that the bits are shifted left instead of right, the result of a shift is to multiply by 2. These properties are illustrated in Figure 3.30 where the number 13 (binary 1101) is loaded. Note that the bit output in the case of division is the remainder.

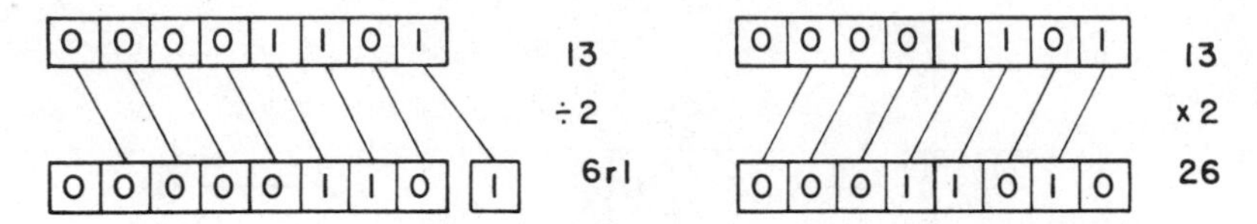

Fig. 3.30 Arithmetic properties of shift registers.

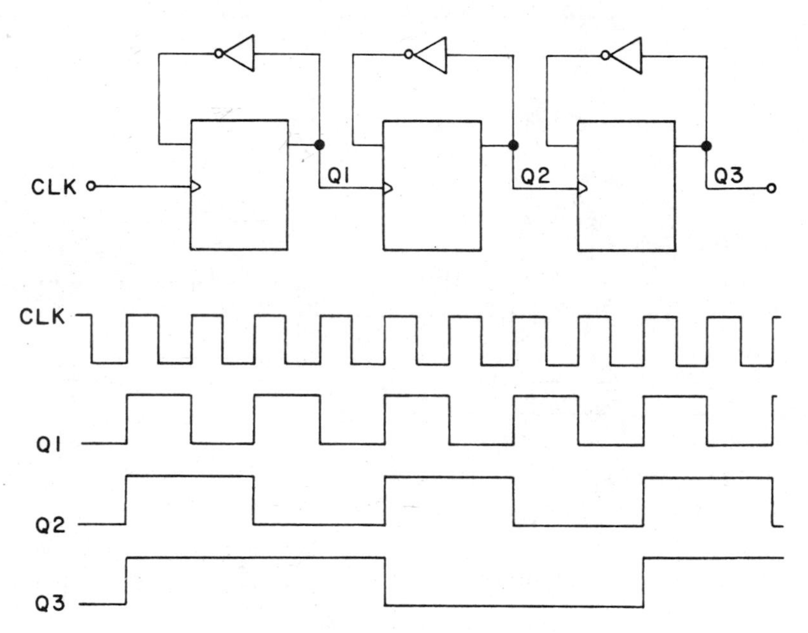

Fig. 3.31 Octal counter circuit diagram.

3.4.1.2 Counters

In the summary of flip-flops we noted that a J–K flip-flop inverts the stored bit if both J and K inputs are held HIGH. The same result can be obtained from a D-type flip-flop if the output is inverted and fed back to the input. In either case the output will be a waveform with half the frequency of the clock input. If this output is used as the clock for a second flip-flop the final signal will have a quarter of the frequency of the original clock, and so on. Figure 3.31 shows an arrangement with three flip-flops and the resulting waveforms. In this example the inverted output drives the clock and the outputs follow the sequence 000, 001, 010, 011, 100, 101, 111, 000, etc. using '0' to represent the output LOW state and '1' the output HIGH state between clock pulses. These are just the binary representations of the numerical sequence 0, 1, 2, 3, 4, 5, 6, 7, 0; in other words, this circuit, if left to itself, will continuously count from 0 to 7 and is called an *octal counter*, octal denoting the fact that there are eight counts in the sequence.

By adding flip-flops we can, apparently, extend the counting capacity indefinitely, but there are two constraints. Firstly, the counts are tied to powers of 2 so that only 2, 4, 8, 16, 32, etc. can be used as the counting base. Secondly, no account has been taken in this analysis of the signal delays through the flip-flops. Every flip-flop which is added to the string will be clocked a little later than the previous one. This restricts the frequency at which the counter can be clocked for the next clock pulse cannot occur until the last flip-flop has had the chance to toggle (change state). This type of counter is called *asynchronous* because all the stages change at different times.

To make a counter in which all the stages change simultaneously, a *synchronous counter*, all the flip-flops must be clocked by the same signal. In this case each stage must tell the following stage when it is to toggle. This may be done with J–K flip-flops. By recalling with J and K both LOW the output does not change; in a simple counter each stage toggles only when all the preceding stages are all HIGH, thus by feeding the J and K with the 'AND-ed' outputs from all previous stages it will toggle at the correct time – see Figure 3.32.

The principle of gating outputs together to select particular numbers is used to make counters with bases other than powers of 2. To see how this is done we need to investigate a more formal way of representing sequential functions, such as those described in this and the previous section.

3.4.2 Formal description of sequential systems

3-4-2-1 State diagrams

We have used the word 'state' somewhat loosely up to now but will define it more accurately in relation to sequential logic systems. The state of a system is a stable set of logic variables held in a set of flip-flops used to define the system. The set of flip-flops is usually called the *state register* although they may be the same as the *output register* which contains the data intended to be the 'result' to the outside world.

This sounds long-winded but may be clarified by looking at examples. An octal synchronous counter, as in Figure 3.32, contains three flip-flops which are both state register and output register. The flip-flops define both the eight stable states of the counter (numbers 0–7) and the useful outputs. The function of the counter is to increment the count by 1 on each clock pulse so the counter must sequence from state '000' to state '001' and then to state '010' and so on. It is

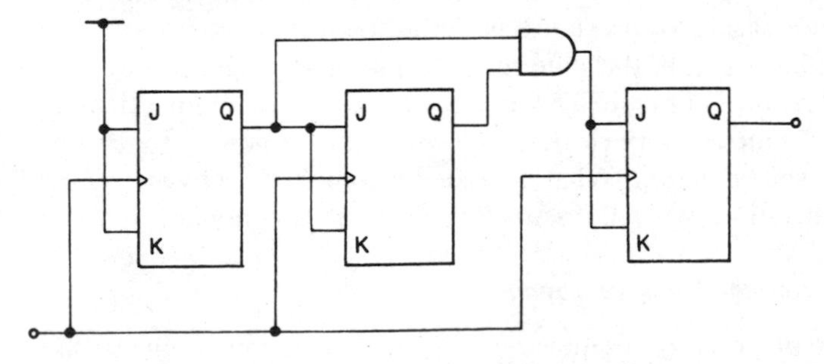

Fig. 3.32 Synchronous counter circuit diagram.

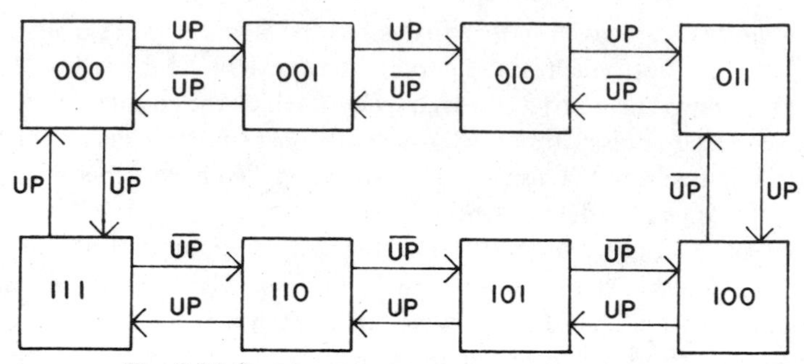

Fig. 3.33 Synchronous counter state diagram.

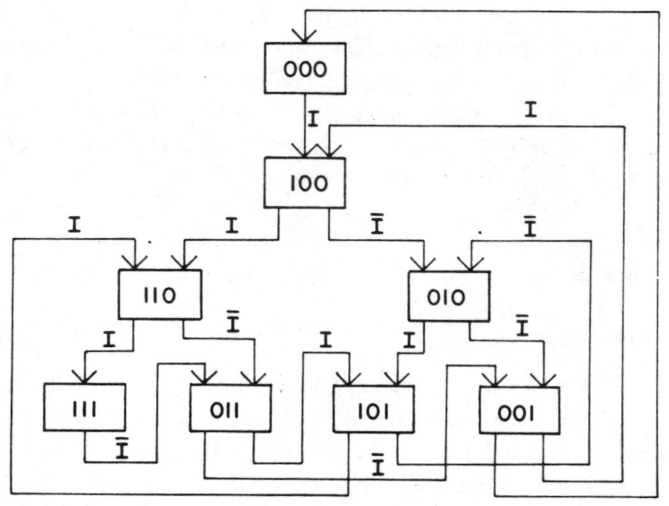

Fig. 3.34 Shift register state diagram.

possible to list all the states and show the progression from state to state by means of arrows. This is the *state diagram* of the system. Figure 3.33 shows the state diagram for the octal counter with one complication; an input signal 'UP' has been included. When UP is HIGH the counter operates as already described, when UP is LOW it counts in the opposite direction; our counter has become an up/down octal synchronous counter.

It is also possible to draw the state diagram for a shift register. In this case all the flip-flops again form a state register but there is only one output flip-flop, the least significant bit of the state register. The state diagram for a three-bit shift register is shown in Figure 3.34; unlike the counter diagram it does not describe the circuit function very clearly, nor would it be very easy to design the circuit from the state diagram. What is needed is a more direct way of describing the circuit operation. We will discuss this in the next section.

3.4.2.2 Karnaugh maps ... again

The state diagram for a sequential system is analogous to the truth table for a combinational system, so we will see how a sequential system can be described

	I	0	0	1	1
	O3	0	1	1	0
O2	O1				
0	0				
0	1				
1	1	H	H	H	H
1	0	H	H	H	H

O3

	I	0	0	1	1
	O3	0	1	1	0
O2	O1				
0	0				
0	1	H	H	H	H
1	1	H	H	H	H
1	0				

O2

	I	0	0	1	1
	O3	0	1	1	0
O2	O1				
0	0			H	H
0	1			H	H
1	1			H	H
1	0			H	H

O1

Fig. 3.35 Karnaugh maps – 3-bit shift register.

with a Karnaugh map. Because the state which the system takes up depends on the state it is currently in as well as the inputs, each bit of the state register has to be treated in the same way as an input. Figure 3.35 shows the maps for each of the three flip-flops in the three-bit shift register; it is clear that O1 goes HIGH when the input is HIGH, irrespective of any other register bits, O2 goes HIGH only when O1 is HIGH and O3 is HIGH when O2 is HIGH. It should be noted that the input conditions refer to the conditions before the clock (the *present state*), while the output is the result of the clock (the *next state*).

While combinational logic is capable of only two output levels there are four possibilities with sequential logic. These are HIGH, LOW, TOGGLE and NO CHANGE. The shift register was adequately defined using only HIGH and LOW, but other circuits may be described better using the alternative

	UP	0	0	1	1
	O3	0	1	1	0
O2	O1				
0	0	H	L	H	L
0	1	L	H	H	L
1	1	L	H	L	H
1	0	L	H	H	L

O3

	UP	0	0	1	1
	O3	0	1	1	0
O2	O1				
0	0	H	H	L	L
0	1	L	L	H	H
1	1	H	H	L	L
1	0	L	L	H	H

O2

	UP	0	0	1	1
	O3	0	1	1	0
O2	O1				
0	0	H	H	H	H
0	1	L	L	L	L
1	1	L	L	L	L
1	0	H	H	H	H

O1

Fig. 3.36 Karnaugh maps – octal counter in terms of 'H' and 'L'.

transitions. For example, the maps for an octal counter using only HIGH and LOW (Figure 3.36), are more complex than the maps using TOGGLE and NO CHANGE (Figure 3.37). This shows that the actual circuit using J–K flip-flops, which allow the TOGGLE and NO CHANGE, will be less complex than the same function built from D-type flip-flops, which allow only HIGH and LOW to be loaded.

3.4.3 Extension of design example

3.4.3.1 Functional description

Having looked at sequential circuit design we are now in a position to extend the

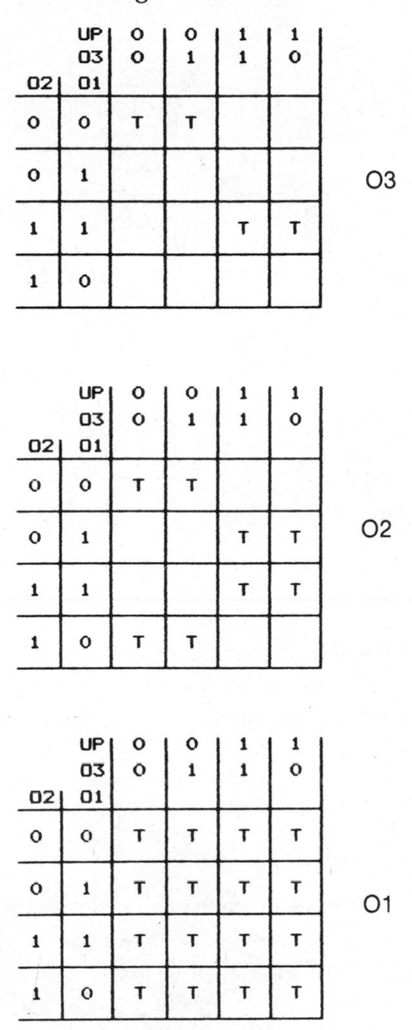

O3

O2	UP / O3 / O1	0 / 0	0 / 1	1 / 1	1 / 0
0	0	T	T		
0	1				
1	1			T	T
1	0				

O2

O2	UP / O3 / O1	0 / 0	0 / 1	1 / 1	1 / 0
0	0	T	T		
0	1			T	T
1	1			T	T
1	0	T	T		

O1

O2	UP / O3 / O1	0 / 0	0 / 1	1 / 1	1 / 0
0	0	T	T	T	T
0	1	T	T	T	T
1	1	T	T	T	T
1	0	T	T	T	T

Fig. 3.37 Karnaugh maps – octal counter in terms of 'toggle' and 'hold'.

function of the combination lock, which we have been using as a design example throughout the book. As a practical system the lock, as described so far, would be rather inadequate for a real security application. After all, by loading only four bits there is a 1 in 16 chance of hitting the correct combination, nor would it take very long to find it by trial and error. By using a sequential system the number of possible combinations can be greatly increased, and incorrect guesses can be detected.

The extended lock will require three digits to be entered in the correct order and only one mistake will be allowed. The system will have four inputs, to allow a BCD number to be entered, plus an 'input' switch to provide a clock signal. The state register will have four bits, two to count the correct sequence, one to detect a mistake and one to signal more than one mistake. The start state has all bits LOW and a correct sequence is detected by the two relevant bits being HIGH.

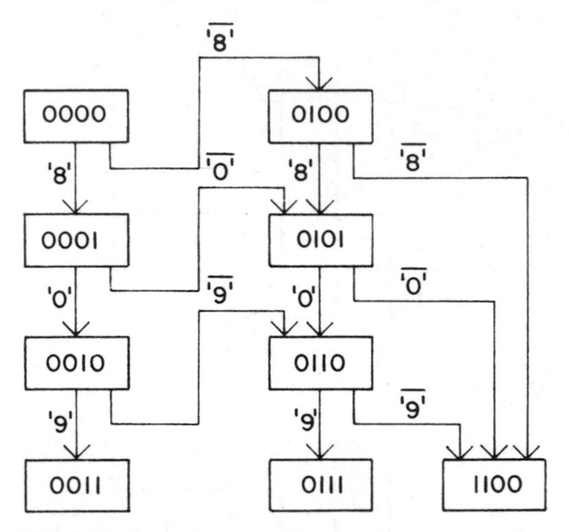

Fig. 3.38 Extended combination lock state diagram.

3.4.3.2 State diagram

This exercise will demonstrate two important practical points regarding state diagrams. The first is that the starting-point must be well defined, usually by ensuring that the state flip-flops take up known levels when power is first applied. In our example I have assumed that all flip-flops power up LOW. The flip-flops in the state register are defined as:

- Q3 – 'alarm'
- Q2 – first mistake
- Q1, Q0 – main sequence

The state diagram is shown in Figure 3.38, starting with '0000' at the top left-hand corner. There are two possible jumps from this state; if the input is '8', the correct number, then Q0 is set as the first step on the sequence. If the input is not '8' then Q2 is set to show that a wrong entry has been made. A sequence of entries '8' – '0' – '9' with no mistakes will cause a progression down the left-hand column until Q1 and Q0 are both HIGH. The same sequence with one mistake will cause a progression down the centre column again ending up in a correct conclusion. A second error will cause a jump from the second column to the bottom right-hand state where Q3, the alarm bit, is set.

The second practical point is that having reached one of the states in the bottom row there must be a way for the system to return to the starting-point, otherwise the lock could only be used once without turning the power off and on again. In this case we have provided another input, RESET, which takes the system back to state '0000' from any of the finish states. From a practical point of view this is a valid function. An authorised user will need to reset the lock after use whilst an unauthorised user will probably not want to wait around if an alarm is set off.

To summarise, a sequential system must have a well-defined entry state and must not include dead ends; every state must have a way in and a way out.

3.4.3.3 Logic equations

In order to construct the system, we need to describe the logic in terms which allow us to decide which gates, flip-flops, etc. are needed, and how they should be connected. We can derive logic equations directly from the state diagram by seeing the input and present state conditions required to set each bit HIGH. Conventionally the symbol ':=' is used to represent the function 'equals after the clock edge'. Equations for the four state bits are then:

$$
\begin{aligned}
Q3 \quad & := Q2 * \overline{Q1} * \overline{Q0} * \overline{(I3 * \overline{I2} * \overline{I1} * \overline{I0})} \\
& + Q2 * \overline{Q1} * Q0 * \overline{(\overline{I3} * \overline{I2} * \overline{I1} * \overline{I0})} \\
& + Q2 * Q1 * \overline{Q0} * \overline{(I3 * \overline{I2} * \overline{I1} * I0)}
\end{aligned}
$$

$$
\begin{aligned}
Q2 \quad & := Q2 \\
& + \overline{Q2} * \overline{Q1} * \overline{Q0} * \overline{(I3 * \overline{I2} * \overline{I1} * \overline{I0})} \\
& + \overline{Q2} * \overline{Q1} * Q0 * \overline{(\overline{I3} * \overline{I2} * \overline{I1} * \overline{I0})} \\
& + \overline{Q2} * Q1 * \overline{Q0} * \overline{(I3 * \overline{I2} * \overline{I1} * I0)}
\end{aligned}
$$

$$
\begin{aligned}
Q1 \quad & := \overline{Q3} * \overline{Q1} * Q0 * \overline{I3} * \overline{I2} * \overline{I1} * \overline{I0} \\
& + \overline{Q3} * Q1 * \overline{Q0} * I3 * \overline{I2} * \overline{I1} * I0
\end{aligned}
$$

$$
\begin{aligned}
Q0 \quad & := \overline{Q3} * \overline{Q1} * \overline{Q0} * I3 * \overline{I2} * \overline{I1} * \overline{I0} \\
& + \overline{Q3} * \overline{Q2} * \overline{Q1} * Q0 * \overline{(\overline{I3} * \overline{I2} * \overline{I1} * \overline{I0})} \\
& + \overline{Q3} * Q1 * \overline{Q0} * I3 * \overline{I2} * \overline{I1} * I0
\end{aligned}
$$

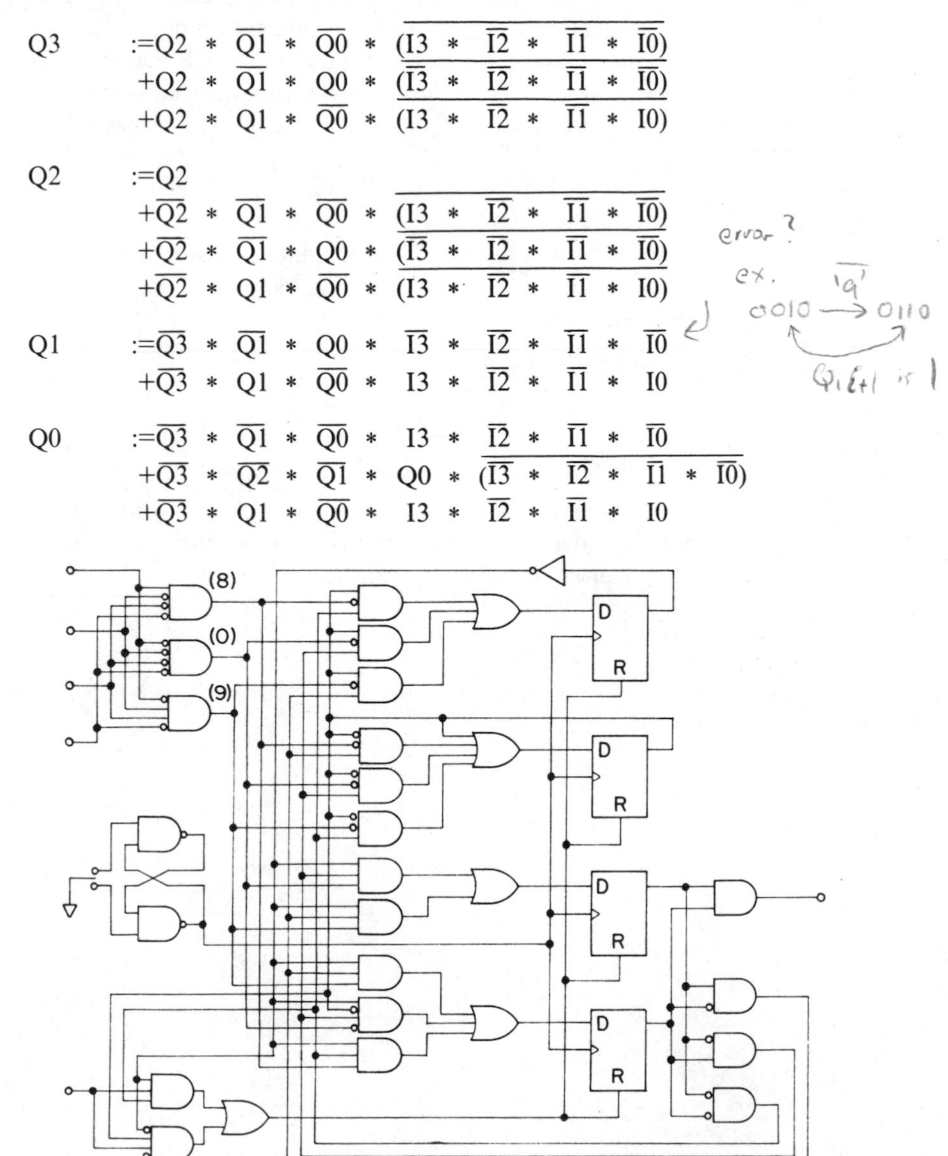

Fig. 3.39 Extended combination lock circuit diagram (discrete logic).

It is assumed that the reset to '0000' will be achieved by an asynchronous reset to the flip-flops. In that case an equation can be written for the reset, if this signal is called 'R':

$$R = RESET * \overline{Q3} * \overline{Q1} * Q0 + RESET * Q3 * Q2 * \overline{Q1} * \overline{Q0}$$

When designing the circuit based on the above equations, it is possible to save hardware by noting that some logic functions, or their complement, are repeated for several of the outputs. For example, the inputs are always decoded to '0' '8' or '9' or their complement, and the state bits Q1 and Q0 are also decoded more than once each in the equations. A possible hardware implementation is shown in Figure 3.39, where an additional output signal, UNLOCK, has been derived as:

$$UNLOCK = Q1 * Q0$$

This is the signal to show that the correct combination has been successfully entered. It also illustrates the use of an R–S latch to 'debounce' the switch input for the clock signal. Altogether the circuit uses:

- twenty-two AND or NAND gates
- five OR gates
- eight inverters
- four D-type flip-flops

These would require about a dozen integrated circuit packages from a standard logic family to construct the circuit. We can now proceed to examine how this can be reduced to one programmable logic device.

Chapter 4
Combinational PLDs

4.1 PROMs

4.1.1 Programmable logic concept

We have seen, in previous chapters, how electronic components can be built into logic components and how these logic components can be built into useful logic systems. In all this discussion it was assumed that the logic components were fixed and unalterable. That is, if we wanted a NAND gate we could buy a small integrated circuit which would have one or more NAND gates inside it, and that we could interconnect several of these packages by copper tracks on a printed wiring board to make the desired logic function. By anticipating the requirements of electronics engineers the integrated circuit manufacturers have also produced more complex functions, by transferring the copper wiring to aluminium tracks on the silicon surface. The designer is still constrained to using those functions which the manufacturer has provided for him.

The only way around this has been in those cases when an engineer has wanted large quantities of the same function, in which case it has been possible for the manufacturer to make him his own circuit by using custom or semi-custom techniques. We have seen one case, however, where the engineer has been able to modify the integrated circuit to meet his own particular requirements. That is where he is using microprocessors and the operating program can be held in a PROM. In examining the structure of a PROM we found that it was built from logic circuits which are used to address a programmable memory matrix. There is no reason then why the PROM itself cannot contain a logic function.

This is the concept of programmable logic. For the remainder of the book we shall develop this concept into practical devices which have the generic name *Programmable Logic Devices*, shortened to PLDs. We shall start by looking at PROMs themselves.

4.1.2 PROMs as logic devices

As we have seen earlier, a convenient way of representing a logic function is by means of a truth table. By way of example we used, in Figure 3.12, a combination lock circuit with two valid combinations of four input bits. We have also noted the convention whereby a HIGH logic level represents the binary digit '1' and a LOW level '0'. Thus the truth table could be written as:

Input	*Output*
9	1
A	1

'A' is the hexadecimal notation for decimal number '10'. Rather than build this circuit out of discrete logic gates it would be just as valid to make it from a PROM. The PROM would be programmed so that addresses '9' and 'A' gave an output of '1' while all other addresses gave an output of '0'.

When looked at in basic terms, there is no difference between a combinational logic device and a memory. Both are designed to give a fixed output in response to a given input combination. The only difference is in their application. A memory has a rather narrow application, particularly, but not exclusively, in storing microprocessor operating programs. Logic devices, of course, have a much wider application.

More relevant than the truth table is the Karnaugh map. Each cell in a Karnaugh map represents a unique combination of input logic levels; indeed, every possible combination is represented by one of the cells. In a PROM every possible combination of input signals will drive one, and only one, row of the memory matrix. Thus, there is a 1-to-1 relationship between map cells and memory cells and it is possible to derive the memory table of a PROM directly from the Karnaugh map of the logic function. In practice, this would be tedious apart from the simplest of functions involving small PROMs, so this procedure is usually carried out by a computer program. Details of software capabilities and availability are discussed in a later chapter.

4.1.3 Practical limitations of PROMs

The above discussion begs the question as to why all logic circuits have not been replaced by PROMs; consider the advantages. All the logic functions you need can be made from one, or at least a small number of PROMs, so buying and stocking are much simpler while the designer is not restricted to set logic functions in a particular integrated circuit. Some reduction in the number of packages is bound to occur as logic systems do not need to be designed with individual gates, but by the relationship between inputs and outputs. If a mistake is made in the design it can probably be corrected by making a small change to the contents of the PROM instead of changing the printed circuit board.

These are all valid points, but there is a price to pay. The complexity of PROMs makes them more costly than simple logic integrated circuits; a 256×8 PROM contains over 100 gates plus a 2048-fuse matrix, so on purely economic grounds a considerable reduction in package count is necessary. The performance of a PROM is also inferior to discrete logic circuits in the same technology. For example, a typical PROM causes about 30 ns of signal delay for a current consumption of over 100 mA (assuming bipolar technology). A discrete logic circuit with the same number of gate delays would cause about

10 ns delay with less than 50 mA. Again, only by condensing the logic into fewer packages does it become viable to design with PROMs. There are areas where PROMs come into their own as we shall see later.

4.1.4 PROM availability

Part of the function of this book is to be a practical guide to the designer of logic circuits so there will be lists showing what is currently available and giving some indication of performance. PROMs can, as we have stated, be used as memories or as logic devices. This is an artificial distinction based on the use to which the PROM is being put, rather than an intrinsic property of the PROM. Nevertheless, a novel method of naming PROMs has been devised by Monolithic Memories, Inc. to emphasise their logic capability. They have coined the term *Programmable Logic Element* or PLE (a trademark of MMI), and describe each PLE as an *mPn*, where *m* is the number of inputs and *n* the number of outputs. The *P* refers to the output level defined in the Karnaugh map. Some other PLD families are restricted to active HIGH outputs or active LOW outputs; clearly, a PLE can have its outputs defined as active HIGH or LOW and is thus classified as having 'programmable' polarity outputs.

Process	*PLE*	*Memory*	*Inps*	*O/Ps*	*Delay*	*Current*
bipolar	5P8	32 x 8	5	8	15 ns	100 mA
bipolar	8P4	256 x 4	8	4	25 ns	120 mA
bipolar	8P8	256 x 8	8	8	30 ns	140 mA
bipolar	9P4	512 x 4	9	4	30 ns	130 mA
bipolar	9P8	512 x 8	9	8	30 ns	155 mA
bipolar	10P4	1024 x 4	10	4	35 ns	140 mA
bipolar	10P8	1024 x 8	10	8	35 ns	170 mA
bipolar	11P4	2048 x 4	11	4	35 ns	150 mA
bipolar	11P8	2048 x 8	11	8	35 ns	175 mA
bipolar	12P4	4096 x 4	12	4	30 ns	155 mA
bipolar	12P8	4096 x 8	12	8	35 ns	175 mA
bipolar	13P8	8192 x 8	13	8	40 ns	175 mA
NMOS	11P8	2048 x 8	11	8	350 ns	60 mA
NMOS	12P8	4096 x 8	12	8	200 ns	100 mA
NMOS	13P8	8192 x 8	13	8	180 ns	75 mA
NMOS	14P8	16384 x 8	14	8	150 ns	100 mA
NMOS	15P8	32768 x 8	15	8	250 ns	100 mA
NMOS	16P8	65536 x 8	16	8	250 ns	140 mA
CMOS	9P8	512 x 8	9	8	150 ns	5 mA
CMOS	10P8	1024 x 8	10	8	30 ns	90 mA

Process	*PLE*	*Memory*	*Inps*	*O/Ps*	*Delay*	*Current*
CMOS	11P8	2048 x 8	11	8	35 ns	90 mA
CMOS	12P8	4096 x 8	12	8	350 ns	5 mA
CMOS	13P8	8192 x 8	13	8	200 ns	10 mA
CMOS	14P8	16384 x 8	14	8	200 ns	20 mA
CMOS	15P8	32768 x 8	15	8	200 ns	30 mA
ECL	8P4	256 x 4	8	4	17 ns	160 mA

Certain conclusions may be drawn from the above table. Delay and power consumption increase as size increases, as might be expected, although there are discontinuities. These are due to improvements in technology which are not applied retrospectively to smaller PROMs, as in general the larger PROMs are the most recent. The greatest activity in new designs is in CMOS although nearly all CMOS, and NMOS, PROMs are used in true memory applications. There will probably not be any significant advances in size of bipolar PROMs because of the problems of handling logic functions with large numbers of inputs, and because of the introduction of alternative methods of coping with high input counts. It is now worth looking at applications where PROMs have been used to advantage over other methods of logic implementation.

4.1.5 PROM applications

4.1.5.1 Address decoding

We have already seen how some microprocessor systems use a multiplexed data and address bus to save pins. The address output is used to find the next instruction in the program, but it is also used to define the location where temporary storage of data can be made. *Random Access Memory* or RAM is used for temporary data storage. A detailed description of RAM circuits is beyond the scope of this book; suffice it to say that as well as being readable in the same way as a ROM, data may be written into each memory location. There is usually an input pin to tell the RAM whether it is read mode or write mode. RAMs and ROMs also have an input called 'output enable'. When this input is inactive the output exhibits a high impedance which is neither HIGH nor LOW. This third output state, sometimes called *tri-state* (a trademark of National Semiconductor Corp), enables the outputs of several devices to be connected together without affecting each other, provided that only one is enabled at any one time. In this way large blocks of ROM or RAM may be built up from smaller units.

In a microprocessor system each small unit is given an address, or range of addresses, for enabling by the processor; thus the address has to be decoded to find which unit is being selected. The decoding may be done with discrete logic devices in a similar manner to the address decoder inside a PROM, but these may often be replaced by the PROM itself. In the system shown in Figure 4.1 the program needs three 8192 × 8 PROMs to store it, there are 16 384 bytes of RAM, and data interfaces to the system, called I/O, are also given addresses so that they can be selected in the same way. The addresses reserved for each device are

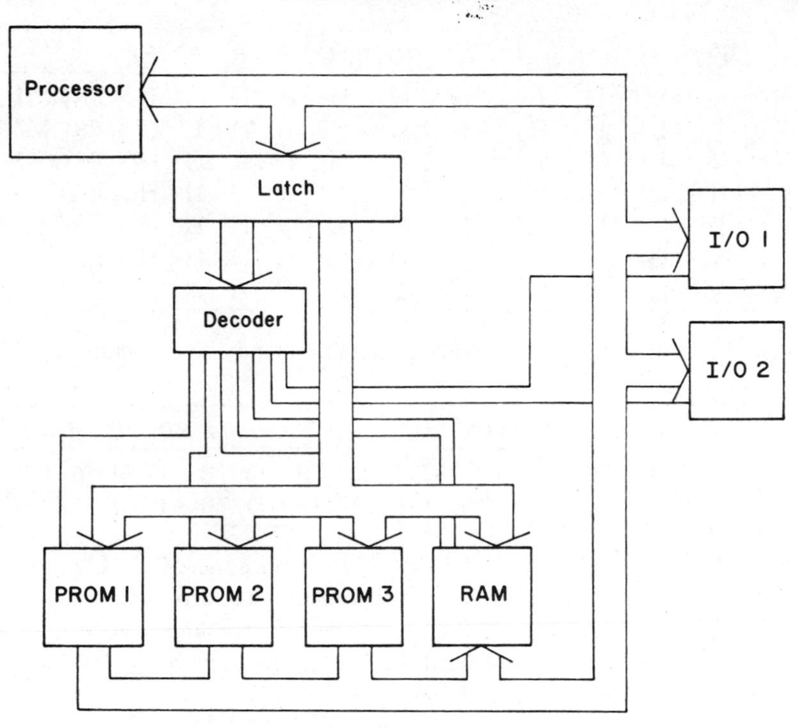

Fig. 4.1 Address decoder used in a microprocessor system.

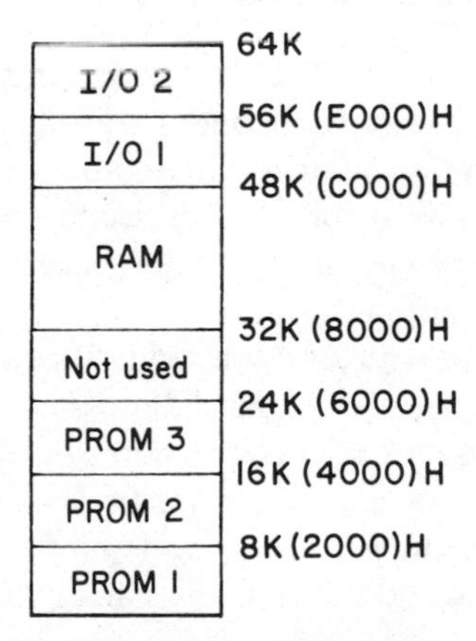

Fig. 4.2 Memory map for microprocessor system.

summarised in a *memory map* which is shown in Figure 4.2. In it, the abbreviation 'k' is used to denote 1024 (2^{10}) bytes.

The smallest block of memory is 8 k (2^{13}) so the lower thirteen address bits are not needed to select any of the blocks. This leaves three address bits plus the signals $\overline{RD}$ and $\overline{WR}$ which are needed for controlling the RAM. The RAM needs two control inputs, the other five units need one each so a total of seven outputs must be supplied from our address decoder PROM. The PLE5P8 comes closest to this configuration with the following truth table:

$\overline{RD}$	$\overline{WR}$	*A15*	*A14*	*A13*	*IO1*	*IO2*	*RAM*	*W*	*P3*	*P2*	*P1*	
X	X	L	L	L	H	H	H	H	H	H	L	- PROM1
X	X	L	L	H	H	H	H	H	H	L	H	- PROM2
X	X	L	H	L	H	H	H	H	L	H	H	- PROM3
L	H	H	L	X	H	H	L	H	H	H	H	- READ RAM
H	L	H	L	X	H	H	L	L	H	H	H	- WRITE RAM
X	X	H	H	L	L	H	H	H	H	H	H	- IO1
X	X	H	H	H	H	L	H	H	H	H	H	- IO2

From this truth table the following address table may be derived:

	0	*1*	*2*	*3*	*4*	*5*	*6*	*7*	*8*	*9*	*A*	*B*	*C*	*D*	*E*	*F*
0000	7E	7D	7B	7F	7F	7F	3F	5F	7E	7D	7B	7F	6F	6F	3F	5F
0010	7E	7D	7B	7F	67	67	3F	5F	7E	7D	7B	7F	7F	7F	3F	5F

The format used above is called *Hex ASCII* because the 'H' & 'L' patterns are converted to the hexadecimal equivalent numbers of address and data. ASCII is the American Standard Code for Information Interchange and is commonly used for sending data between electronic equipments. The active level for all the enable inputs is assumed to be LOW.

4.1.5.2 Code converters and look-up tables

A particularly suitable use of PROMs is for a device known as a look-up table. Mathematical tables of logarithms, trigonometric functions, etc. are well known, their use being to convert one number into another number which is a function of it. Although it is usually possible to devise an algorithm to calculate the converted number, this is often a fairly lengthy process. The alternative is to load a PROM with this information.

The PROM table is constructed by making the address correspond to the number being converted and the output data to the 'answer'. This is illustrated in Figure 4.3. A particularly useful conversion is 'twos complement'; addition and subtraction are two simple arithmetic functions which are common in electronic systems, as in other areas. The logic for these two operations is substantially different, but addition of the twos complement of a number is

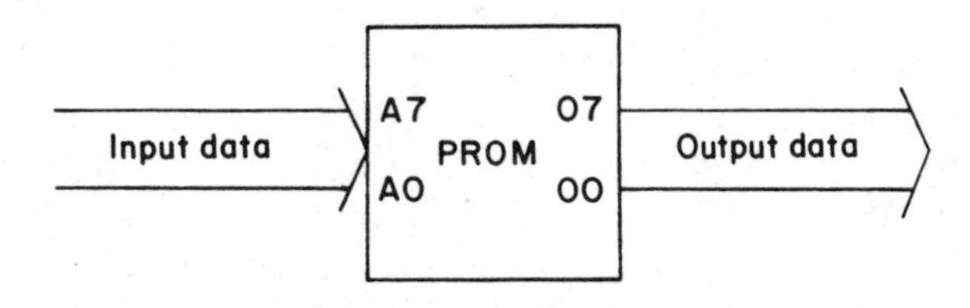

Fig. 4.3 PROM look-up table application.

equivalent to subtracting that number. Thus, if this function is readily available, the same circuitry may be used for subtraction and addition.

A code converter is just a specific example of a look-up table. Converting between decimal numbers and binary numbers is quite tedious but can be done very readily by a PROM. In all these cases it would be possible to work out the equivalent logic of the function being carried out by the PROM, and then create the function out of pure logic circuits. This will normally result in a far more complex solution.

4.2 PLAs

4.2.1 Relationship with PROMs

The last section described how PROMs can be used to implement logic functions. Some of the practical limitations were discussed and we can now explore that subject further. The most notable feature of a PROM, when viewed as a logic device, is that every possible combination of input is present in the memory matrix. This makes PROMs very versatile but has some unpleasant consequences. Firstly, it means that there is often a high redundancy in the programmed information. In the address decoder example the logic function was described in seven lines of truth table, yet there were thirty-two lines of data programmed in the PROM. Secondly, adding an input to a PROM causes the size of the memory matrix to be doubled. This will have repercussions on cost and, to a lesser extent, on performance. Lastly, PROMs come with either four or eight outputs but logic systems often do not; thus output pins often remain unused, as one did in the address decoder example.

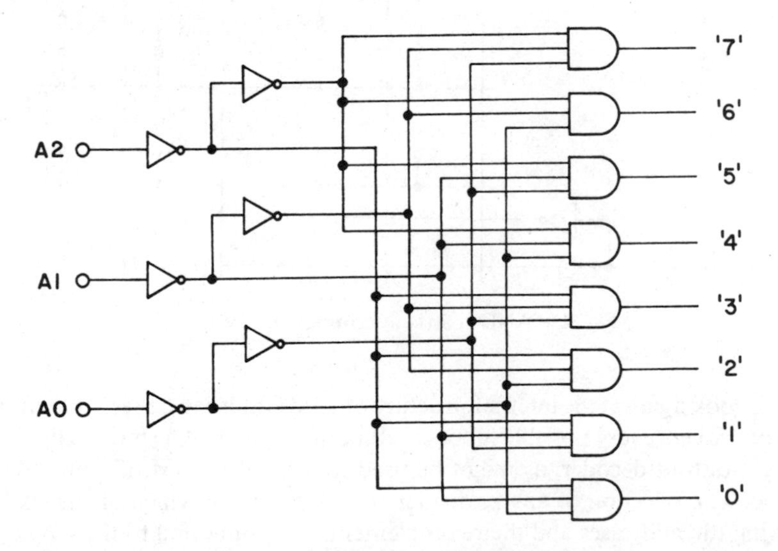

Fig. 4.4 PROM input address decoder.

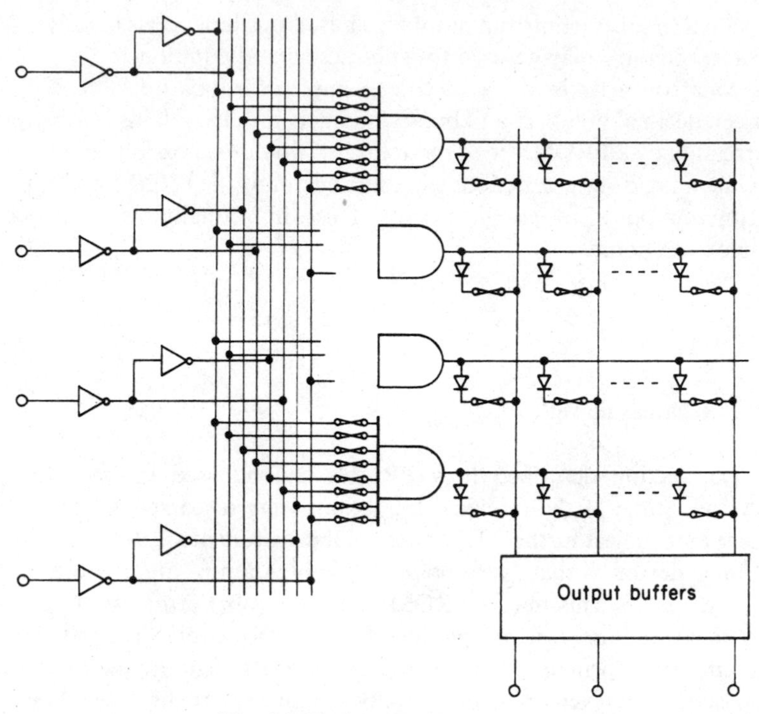

Fig. 4.5 Typical PLA structure.

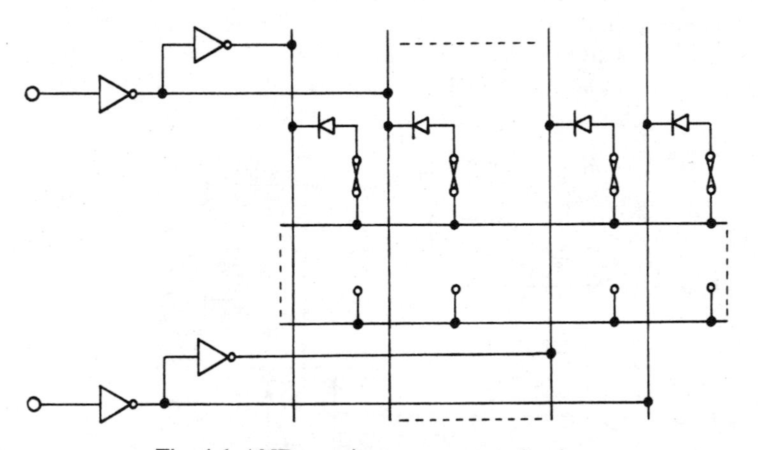

Fig. 4.6 AND matrix at component level.

If we look again at the internal structure of a PROM it can be seen that the row decoder covers every possible input combination. Figure 4.4 shows a three-line to eight-output decoder, as might be used in a 8×4 PROM, if such a device existed. The selection of any particular row depends on which of the six lines carrying the addresses and their complements are connected to the AND gate driving that row. In a *Programmable Logic Array*, or PLA, these connections are

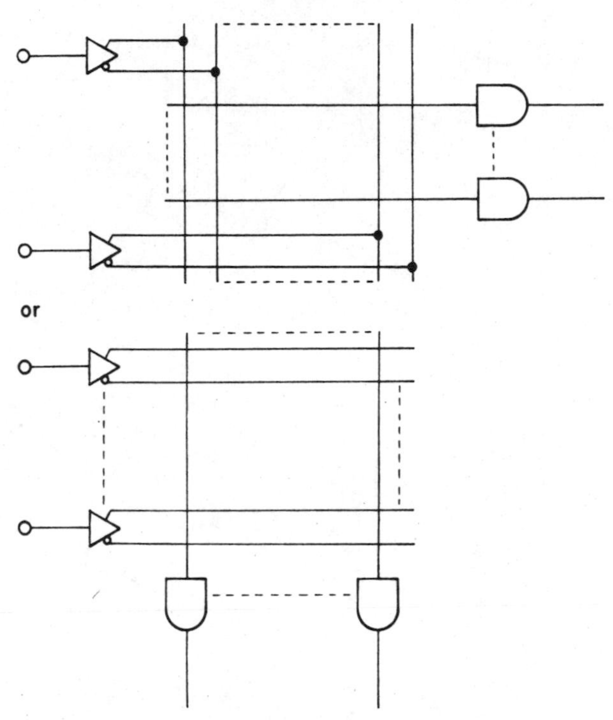

Fig. 4.7 Conventional representation of AND matrix.

made programmable and the number of AND gates is reduced. A typical structure is shown in Figure 4.5.

A PLA then has two programmable sections: the part described above, which connects the inputs to the AND gates is called the *AND matrix*. The second part, already described as part of the PROM structure, determines which AND gates are OR-ed together to form the output function; it is called the *OR matrix*. PLAs differ from PROMs by having a programmable AND matrix, although both have a programmable OR matrix.

4.2.2 Designing with PLAs

4.2.2.1 Configuring the AND matrix

An AND gate can be made from a group of diodes connected by their anodes, so a component level structural diagram of a simple AND matrix is shown in Figure 4.6, the fuse indicating that the matrix is field-programmable. For simplicity this is often drawn as illustrated in Figure 4.7, it being understood that there is a diode and a fuse at each crossover. It is possible to map logic equations directly on to this matrix diagram.

Conventionally, an intact fuse is indicated by a 'X' or '.' at the crossover. Unprogrammed FPLAs have all fuses intact and should therefore have an 'X' at

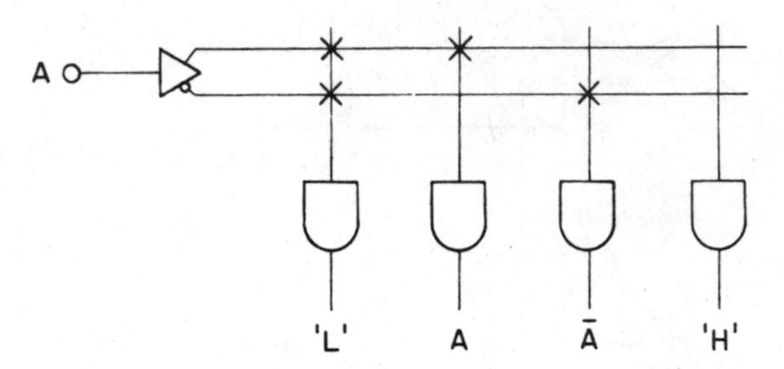

Fig. 4.8 Input connections to AND matrix.

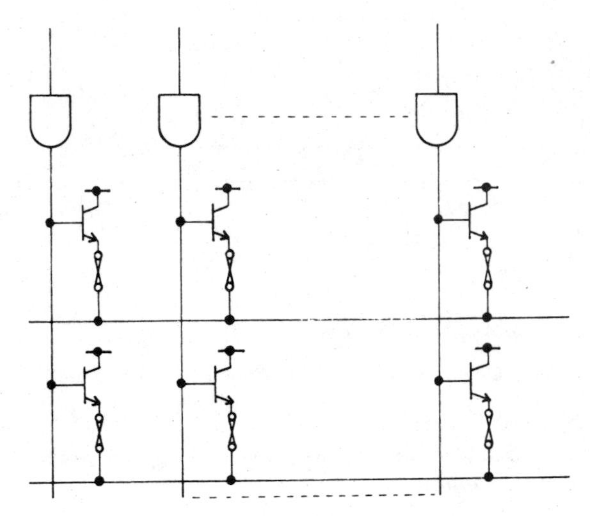

Fig. 4.9 OR matrix at component level.

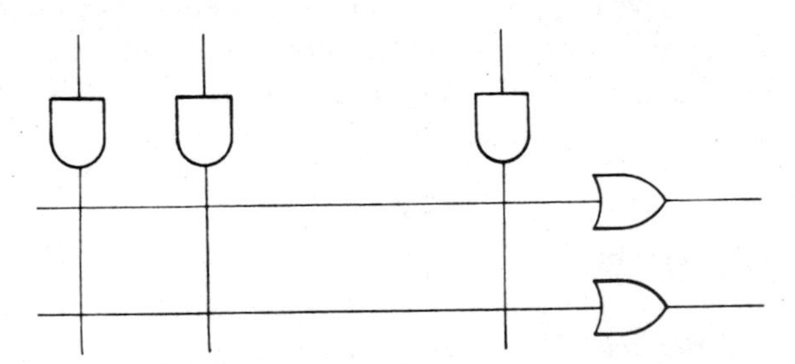

Fig. 4.10 Conventional representation of OR matrix.

every crossover; this could be confusing, so they may be left out completely, with an 'X' in the AND gate symbol indicating an unprogrammed gate.

There are four possibilities for each input to be connected to each gate of the AND matrix, as shown in Figure 4.8. If both an input and its inverted signal are connected, we have the logic statement $A*\bar{A}$ which is always LOW, therefore any gate with this included will inevitably be inactive. With the complement fuse blown, the true sense of the input (A) is included in the equation for that gate. If the inverted input is connected the complement ($\bar{A}$) will appear in the equation. When neither the input nor its complement is connected, that signal does not appear at all in the equation for that gate; this is the 'don't care' condition. If all the inputs are disconnected from an AND gate its output will be permanently HIGH, as may be seen from Figure 4.6.

4.2.2.2 Configuring the OR matrix

The OR matrix of a PLA is no different from that of a PROM. Figure 4.9 shows the component level structure and Figure 4.10 the conventional way of representing this. Each crossover is presumed to contain a fuse and a diode, connected by its cathode to the other diodes in the same gate. Again a 'X' or '.' is used to show that a signal from the AND matrix is connected to that OR gate and, therefore, included in that output function. It should be clear from the structural diagram that any OR gate with all its inputs disconnected will have a permanently LOW output.

4.2.3 Formal PLA methods

4.2.3.1 Truth table entry

We saw in the previous chapter that logic functions may be conveniently described by a truth table. This may be used as the basis for entering the information needed to program a PLA. Most commercial programming equipment will accept a truth table as a direct entry method for PLAs, but it is worth while pointing out the relationship between the truth table data and the fuses which are blown from that data. An 'H' in the truth table is equivalent to a true input, so this requires the complement fuse to be blown leaving the non-inverting connection intact. Conversely, an 'L' is interpreted by blowing the true fuse so that only the inverted input is connected.

The third possible entry in the truth table is '–'; this is the condition where the input is not included at all, so both fuses are blown in this case. The fourth possibility, both fuses being left intact, has the effect of disabling the gate to which they are connected, as we saw above. This condition is not normally specified in PLA logic tables; if it is required the symbol '0' is used.

We also need to specify whether the AND gate is connected to the output via the OR matrix, or not. An intact fuse means that the connection is made and this is shown by an 'A' in the appropriate output column. An AND gate which is not included in an output must have the fuse blown in the OR matrix; this is indicated by a '.' in the output column.

4.2.3.2 Logic equation entry

Logic equations may be used as a source of programming data in one of two ways. Computer assistance for using logic equations is described in Chapter 8; it is also possible to convert equations to a truth table, by hand, and use this as the basis for data entry. True signals in the equation are entered into the table as 'H' and complemented signals as 'L'. Any signal not appearing in the equation is entered as '–'. Each line in the table corresponds to a single AND gate so the equations must be arranged into AND–OR format by using the rules specified in Chapter 3.

The group of signals connected into a single AND gate is called an *AND term*. If the same AND term appears in more than one equation, it need be specified only once and connected to the necessary outputs by an 'A' in the appropriate column. We can illustrate the conversion process by looking at the address map example from Section 4.1.5, as below:

$\overline{RD}$	$\overline{WR}$	*A15*	*A14*	*A13*	*IO1*	*IO2*	*RAM*	$\overline{W}$	*P3*	*P2*	*P1*
–	–	L	L	L	.	.	.	.	.	.	A
–	–	L	L	H	.	.	.	.	.	A	.
–	–	L	H	L	.	.	.	.	A	.	.
L	H	H	L	–	.	.	A	.	.	.	.
H	L	H	L	–	.	.	A	A	.	.	.
–	–	H	H	L	A	.	.	.	.	.	.
–	–	H	H	H	.	A	.	.	.	.	.

This table may now be entered directly into a programmer, and a PLA produced with that logic function.

4.2.3.3 Karnaugh map analysis

We have already noted that the PROM structure contains every cell in the Karnaugh map, so it is possible to implement any combinational logic function in a PROM, provided that there are enough inputs. The power of the PLA is that inputs can be added without needing to increase the number of AND gates. There is a price to pay for this and that is that there may not be enough AND gates to implement a given function. By mapping functions on to a Karnaugh map it may be possible to arrange the logic in such a way as to reduce the number of gates needed for the function.

It is not always possible to find a meaningful example to illustrate features which tend to occur in random logic situations. Some idea of what can be done is seen in the following example, although more obvious situations are likely to arise in real designs. The circuit in question has an output which equals the number of HIGHs on the input lines, as is shown in the following truth table:

I3	*I2*	*I1*	*I0*	*O2*	*O1*	*O0*	*I3*	*I2*	*I1*	*I0*	*O2*	*O1*	*O0*
0	0	0	0	0	0	0	1	0	0	0	0	0	1
0	0	0	1	0	0	1	1	0	0	1	0	1	0
0	0	1	0	0	0	1	1	0	1	0	0	1	0

I3	*I2*	*I1*	*I0*	*O2*	*O1*	*O0*	*I3*	*I2*	*I1*	*I0*	*O2*	*O1*	*O0*
0	0	1	1	0	1	0	1	0	1	1	0	1	1
0	1	0	0	0	0	1	1	1	0	0	0	1	0
0	1	0	1	0	1	0	1	1	0	1	0	1	1
0	1	1	0	0	1	0	1	1	1	0	0	1	1
0	1	1	1	0	1	1	1	1	1	1	1	0	0

The Karnaugh maps for the three output signals are shown in Figure 4.11. If this function is implemented in a PLA with no output inversions then fifteen AND gates are needed, only one less than using a PROM! O0 needs eight gates, O1 six gates and O0 one gate. Even if the four cells which are 'H' in both O0 and

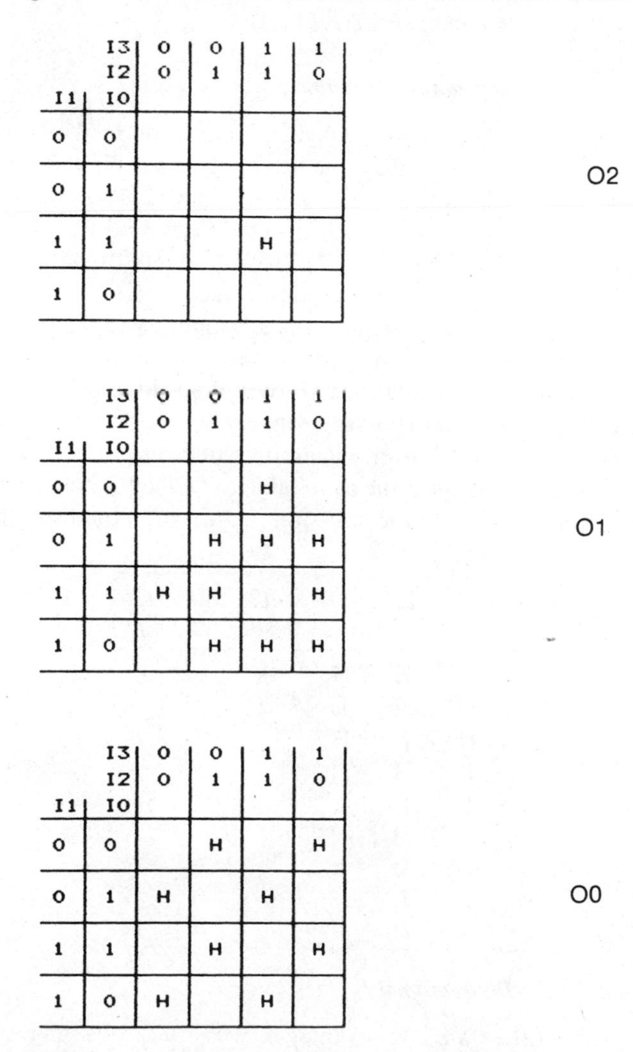

Fig. 4.11 Karnaugh maps – bit counting circuit.

O1 are shared, a further six gates are still required to complete O1. However, if O1 is inverted, one of its AND gates can be used for O2, four can be used in O0 and five more are needed to complete O0 and O1, a total of ten altogether.

This is not a startling improvement, partly due to the fact that we have limited the function size to four inputs for the sake of simplicity, but it illustrates the principle of logic minimisation in PLAs, that is, to minimise the total number of gates by arranging for outputs to share gates wherever possible. In this example all the gates would use all four inputs; in a discrete logic solution that might not be optimal for four-input gates take half a TTL or CMOS package and a solution using two-input and three-input gates may use fewer packages.

4.2.4 PLA output structures

4.2.4.1 Programmable output polarity

In order to be able to use logic minimisation methods without adding external components, most PLAs have a programmable inverter built into the output structure. The way in which this is achieved is shown in Figure 4.12. One input to an exclusive-OR gate is connected to ground via a fuse; referring back to the truth table in Section 3.3.1.2, it is clear that with this fuse intact, that is with one input always LOW, the output level is the same as the input level. If the fuse is blown an internal resistor takes this input to a HIGH level and the output level is now the complement of the input.

In its non-inverting state the output is said to be *active-HIGH* because the OR gate output is left in the same sense and will only be HIGH when one of the AND gates is active. Conversely, an inverting output is called *active-LOW*. There is usually provision in the truth table to define the active level of each output. The full truth table for the 'bit counting' circuit would thus be:

I3	*I2*	*I1*	*I0*	*O2*	*O1*	*O0*
Active level				*H*	*L*	*H*
H	H	H	H	A	A	.
H	H	H	L	.	.	A
H	H	L	H	.	.	A
H	L	H	H	.	.	A
H	L	L	L	.	A	A
L	H	H	H	.	.	A
L	H	L	L	.	A	A
L	L	L	H	.	A	A
L	L	H	L	.	A	A

4.2.4.2 Bidirectional I/O

One of the stated disadvantages of PROMs was the restriction to four or eight outputs, which can prove wasteful of device pins. So far, we have not addressed

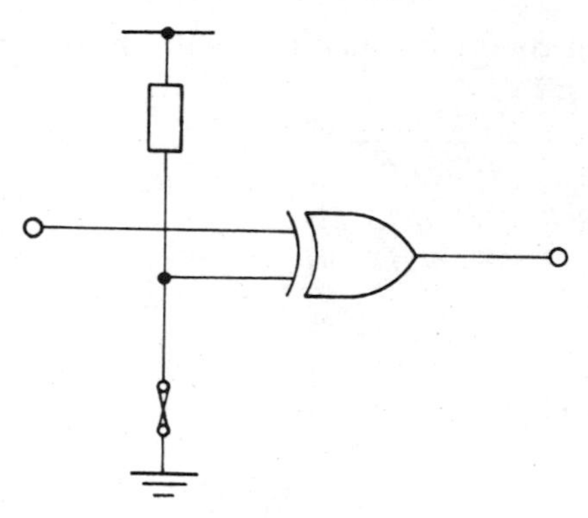

Fig. 4.12 Programmable output inverter.

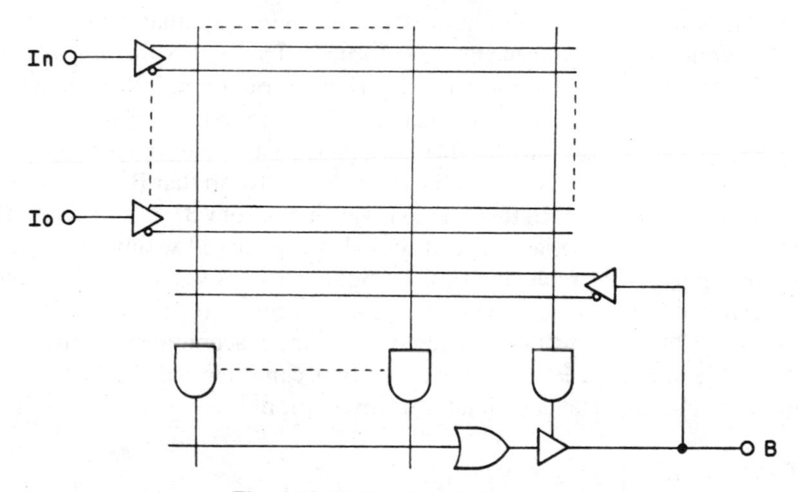

Fig. 4.13 Bidirectional I/O structure.

this problem in relation to PLAs. By making use of the tri-state output structure, described earlier, some of the pins can be arranged to function either as inputs or outputs. This facility is illustrated in Figure 4.13. One AND gate is dedicated to controlling the tri-state output, while the same pin is also connected to the AND matrix as an input. There are four options in the way in which this pin may now be used, depending on how the AND gate is programmed; these are:

- dedicated input pin – all fuses left intact
- dedicated output pin – all fuses blown
- controlled output pin – gate programmed with control logic
- output with feedback – output signal used in AND matrix

The simple D-latch, from Section 3.3.2.2, is a good example of a circuit which can be built from a PLA with bidirectional pins. The control gates are often called *direction gates* because they define signal direction of the pin. In the truth table gate 'Dn' controls pin 'Bn'. Because they drive the output control directly there are no output conditions associated with the direction gates. The truth table of a D-latch is:

	D	*LE*	*Q*	*D*	*LE*	*Q*
	BI3	*BI2*	*BI1*	*BO3*	*BO2*	*BO1*
	Active level			*H*	*H*	*H*
	H	H	–	.	.	A
	–	L	H	.	.	A
	H	–	H	.	.	A
D3	0	0	0			
D2	0	0	0	.		
D1	–	–	–			

This is one occasion when the symbol '0' is used to indicate that the fuses from both the input and its complement are left intact. This makes the output from that gate LOW which puts the output in its tri-state condition. It then has a high impedance, behaving like an open switch, and does not affect any input signals connected to that pin. D1, on the other hand, has all its inputs open which gives a HIGH on the output; this cancels the tri-state control so that B1 behaves just as any normal output pin. In this example the signal from B1 is fed back to the AND matrix to form the latch circuit, but this is achieved without using any other input pins. This circuit could still be built in a PLA without bidirectional pins but in that case it would occupy four pins instead of the three used above.

Thus, as well as allowing the designer to define the exact number of inputs and outputs he requires in a PLA, bidirectional pins can often reduce the total pin count needed for a particular circuit implementation.

4.2.5 PLA availability

Most PLAs are sold under part numbers which bear no relationship to their internal structure, unlike PLEs which we investigated earlier. Also, at the time of writing, PLAs are available only in bipolar technology although we understand that plans are in hand to produce the same structures in CMOS. Current availabilty is as follows then:

Part number	*Input pins*	*B pins*	*Output pins*	*AND gates*	*Prop. delay*	*Supply current*
PLS100	16		8	48	50 ns	170 mA
PLS153	8	10		32	30 ns	155 mA
PLS161	12		8	48	50 ns	170 mA
PLS173	12	10		32	30 ns	170 mA

There are thus just four PLA devices available although, with their flexibility in terms of input/output I/O allocation and the number of gates available for sharing among outputs, they offer at least as much scope for logic designers as PLEs. The full circuit diagram of the PLS153 is shown in Figure 4.14.

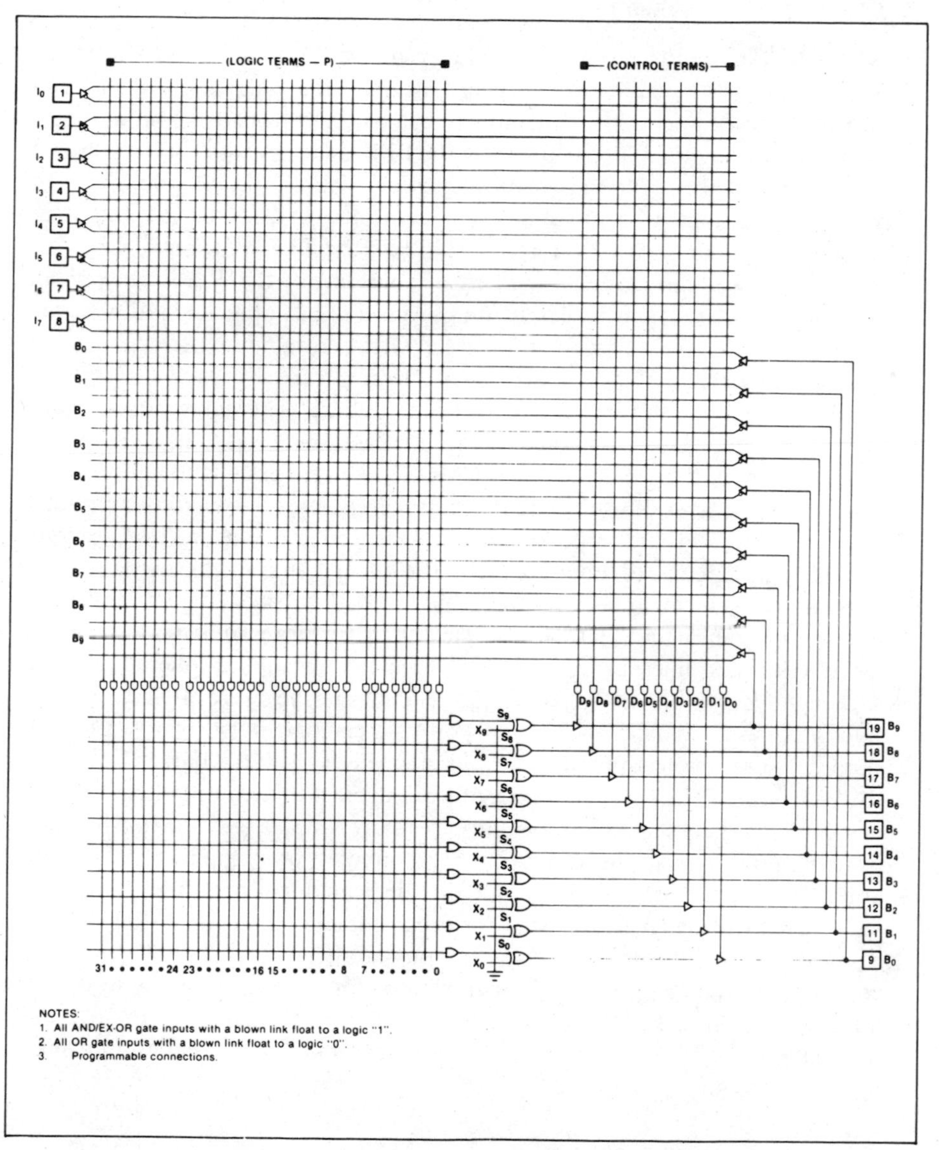

Fig. 4.14 PLS153 circuit diagram. *(Reproduced by permission of Mullard Ltd)*

4.3 PALs

4.3.1 PAL architecture

4.3.1.1 Derivation from PLAs

Where PLA stands for Programmable Logic Array, PAL (a trademark of MMI) stands for Programmable Array Logic. This may seem an artificial distinction introduced to distinguish one manufacturer's product from an existing range, but it did mark a radical departure from the PLE and PLA concepts established during the 1970s. If we take an historical look at the development of programmable logic we can see how this came about. The first PROMs were introduced in about 1970 followed by a PLA, the PLS100, four years later. Although it was an extremely powerful logic device, the PLA was not an immediate success. It was probably too powerful to be appreciated fully by the designers of the time. Also, its inflexible I/O structure limited the number of applications where it could replace discrete logic effectively.

Microprocessors, also introduced at about this time, were supposed to replace the discrete circuits used previously in logic designs; there was to be a revolution which would cut the use of TTL and CMOS integrated circuits to a fraction of their former number. It was found that sales of discrete logic circuits continued to grow in spite of all the predictions. The reason was that microprocessors needed logic circuits to interface with the other components used in their systems. Discrete logic became the 'glue' needed to stick systems together. Designers began to find that the glue was taking up more printed circuit board space than the microprocessors and their intelligent peripheral circuits. Although PROMs and PLAs could replace glue logic they were often an overkill solution, inasmuch as a large proportion of their logic power was wasted. A simpler solution was required and PALs stepped-in to fill the bill.

Much of the PLA's power is derived from the fact that common AND gates can be shared between any of the outputs. Many simple logic functions do not overlap in this manner, however, so it is an unnecessary complication in many applications. In a PAL the AND gates are dedicated to a particular output by making the OR matrix fixed instead of programmable. Figure 4.15 shows how a simple PAL with four inputs, two outputs and eight AND gates would be constructed, while Figure 4.16 shows how the same effect can be obtained with a PLA. Restricting the OR matrix to fixed connections means that some of the flexibility of the PLA is lost, but this is compensated for by it being simpler to use.

4.3.1.2 PAL output structures

Although they have lost the programmable OR matrix, PALs do not suffer fiom any other drawbacks from the architecture standpoint. Indeed, their simplicity makes them attractive to use for replacing discrete logic circuits.

The first PALs to be produced had an inflexible architecture. The outputs were fixed in number and in output polarity, and had few AND gates to share between

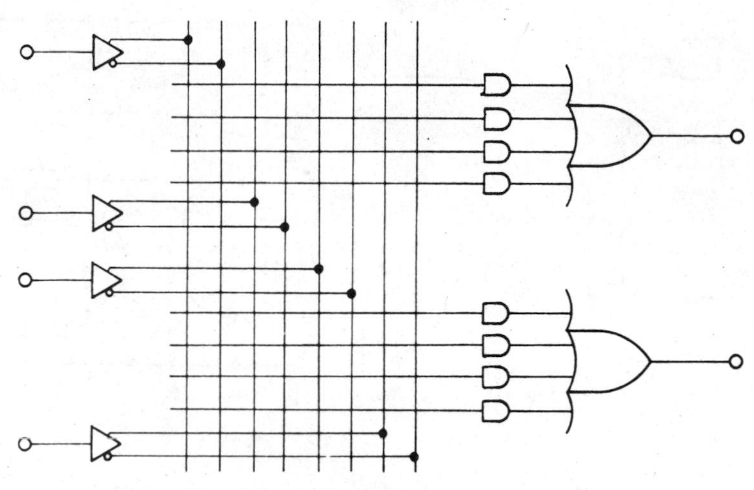

Fig. 4.15 Simplified PAL structure.

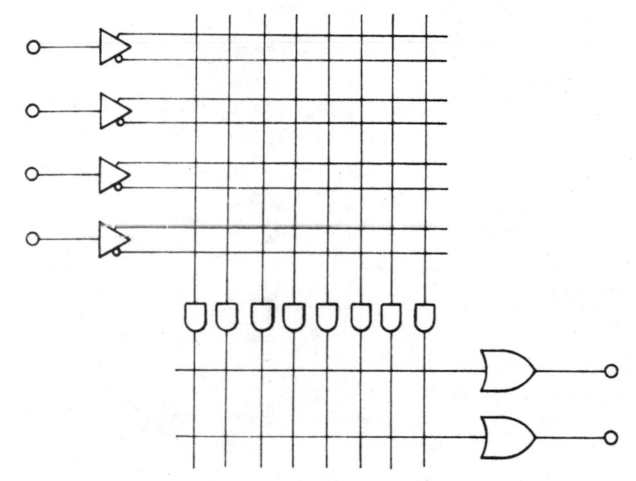

Fig. 4.16 PLA structure for comparison with PAL.

them. This was fine for some applications – such as address decoders where the enabling inputs of the selected devices were all active-LOW – but could result in problems in other cases. Most of the PAL combinations are now available with programmable output polarity, just as the PLAs are. The other structure featured in some PLAs, bidirectional I/O pins, can also be found in some PALs.

While PALs are very attractive for designing less complex circuits they do suffer from the limitation of restricting the number of AND gates which can be ORed together in each output. Even fairly complex PALs have only seven or eight gates per output. One approach to alleviating this problem has been the use of shared AND gates. Typically sixteen AND gates can be shared between two outputs, thus a complex function and a simple function may occupy two

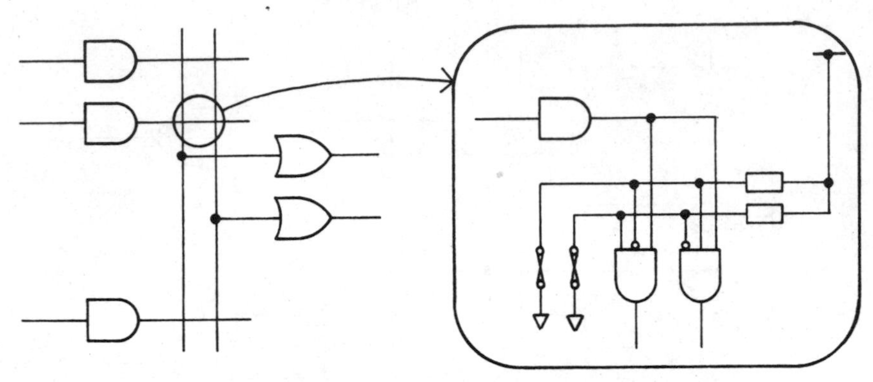

Fig. 4.17 PAL shared output structure.

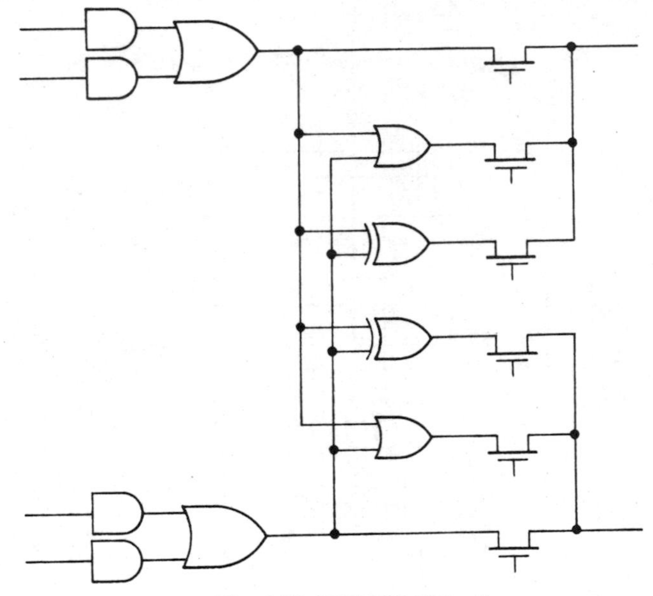

Fig. 4.18 'XPANDOR' cell.

adjacent outputs provided that they need no more than sixteen gates. The shared output structure is illustrated in Figure 4.17.

A recent enhancement of this approach may be said to bridge the gap between PALs and PLAs; this is the so-called Xpandor Cell' (Xpandor being a trademark of VTI). In the shared output structure each AND gate can be allocated to only one of the outputs sharing it; in the Xpandor Cell two outputs can both use any of the two sets of gates allocated to that pair. The OR matrix is not programmable so it is not a true PLA, and if the primary set of AND gates for one of the outputs is used for that output, then that function must be used in full as the secondary set for the alternative output. Figure 4.18 should make that clear.

It also shows that the secondary set of gates may be combined with the first set as either an OR function or an exclusive-OR. The second option can lead to some interesting possibilities whereby more complex functions can be accommodated than at first realised. For example, looking back to Figure 4.11,the Karnaugh map for Q0 shows that eight AND gates would be needed for this function, which is just the exclusive-OR of all four inputs. This function may be realised with only four gates if an exclusive-OR is available. Two gates in each set can form the exclusive-OR of two pairs of the inputs; if the resulting functions are exclusive-ORed again in an Xpandor cell, the required function is produced.

4.3.2 Using PALs

4.3.2.1 Design methods

There is very little difference between designing PLAs and designing PALs. It comes down to establishing a method for describing the logic system in such a way that the appropriate fuses can be blown in whichever device is selected. The tools for assisting the designer will be covered in Chapter 8; here we are more concerned with seeing how the logic can be fitted into the various devices. It is the OR matrix which really governs the way in which logic minimisation is handled, and the Karnaugh map is the key to this.

The programmable OR matrix of the PLA made it politic to look for common AND functions in the maps of all the outputs. This approach will not help at all in PALs because the AND gates are allocated to a single output and cannot be shared, except as described in the previous section. Thus the art in designing for PALs is to reduce the number of AND gates for each output. To fit the example of Figure 4.11 into a PAL would therefore require a PAL with the following properties:

Q2 – active HIGH output with one AND gate, or
  – active LOW output with four AND gates
Q1 – active HIGH output with six AND gates, or
  – active LOW output with six AND gates
Q0 – active HIGH output with eight AND gates, or
  – active LOW output with eight AND gates, or
  – Xpandor cell with 2 × 2 AND gates

The final choice of which device to use will depend on a number of factors, such as speed, power consumption and cost; the steps which need to be taken to achieve this will be discussed later in the book. The way in which the design information is converted to a finished device depend on the design tools which are available.

4.3.2.2 PAL nomenclature

Unlike PLAs, it is possible to deduce much about the structure of any PAL from its part number. PALs are numbered as PAL*mXn*, where *m* is the number of inputs to the AND gate matrix, n is the number of outputs and X defines the output structure. The common letters used for combinational PALs are:

H – active HIGH output
L – active LOW output
P – programmable polarity
S – shared AND gates

Most PALS are supplied as either twenty-pin or twenty-four-pin devices having, respectively, eighteen or twenty-two logic connections. Thus, if $m+n$ comes to either eighteen or twenty-two the device probably does not have bidirectional I/O pins and m is the number of device inputs, otherwise the number of inputs can be guessed by subtracting n from 18 or 22. An additional clue to the structure of a PAL comes from the fact that most PALs can be grouped into two classes of complexity, the simpler PALs having sixteen or twenty AND gates in total with no bidirectional I/O, the others having sixty-four or eighty AND gates and bidirectional I/O. AND gates are normally spread evenly among the outputs so the number of AND gates per output can often be deduced.

4.3.2.3 Security fuse

A feature to be found in PALs but not, as yet, in PLAs is the *security fuse*. While the primary purpose of a PLD is to perform logic functions with fewer packages, there is a secondary benefit which could be even more useful in certain circumstances. The logic function contained in a PLD depends on the fuses which have been blown inside the device. While we have not yet discussed in detail the programming of PLAs and PALs, the methods are very similar to those for blowing PROMs. In other words, they rely on internal circuitry which is in place for allowing the device to be programmed, and for reading which fuses have been blown.

Once a device has been programmed this extra circuitry is redundant. If it is disabled then it becomes impossible to read back the contents of the PLD and therefore its function. This makes it impossible, or at least very difficult, to copy. The security fuse is designed for just this purpose; once it has been blown in a PAL, that PAL cannot be forced to divulge its contents. It therefore becomes the ideal device to use in circuits which the designer wants to make difficult to copy, for commercial or other reasons. In some industries PALs are used to hold secret combinations, or to scramble signal paths, in order to make designs secure, so that competitors cannot make an identical equipment without having to invest themselves in design effort.

4.3.3 PAL availability

4.3.3.1 PAL families

As we stated above, PALs may be grouped in families according to their complexity. There are also PALs available within each group with different speed and power ratings, and different technologies. In our survey of combinational PALs we can therefore split the tables into architecture and performance. Low-complexity PALs are those with sixteen or twenty AND gates, medium complexity PALs have sixty-four or eighty gates.

PALs are fabricated in both bipolar and CMOS technologies but, in order to provide some choice of performance, bipolar PALs are available in different speed and power combinations. Because they have to cater for all their features to be fully used, the power consumption of both PALs and PLAs is often higher than the same circuit constructed from discrete logic. Just as discrete logic families are made with different levels of power consumption (e.g. ALS, AS and FAST), some ranges of PAL have been designed with lower supply currents than the standard family. This gives the designer the option of choosing devices with high speed, or low power if high speed is not required for the application. Families of half-power and quarter-power PALs are available.

The other option for low power is to use CMOS. The current consumption of CMOS is significantly lower at low speeds, but is comparable to bipolar devices at frequencies of about 10 MHz, or higher. This is because the internal nodes have to be charged and discharged more rapidly.

4.3.3.2 Combinational PAL summary

In the following tables, the various output configurations for a given I/O and gating architecture have been included on a single line:

Part Number	*Inputs*	*B I/O*	*Outputs*	*AND gates*
Low complexity PALs:				
10L8 10H8 10P8	10		8	16
12L6 12H6 12P6	12		6	16
14L4 14H4 14P4	14		4	16
16L2 16H2 16P2	16		2	16
16C1	16		2 (Q & $\overline{Q}$)	16
12L10 12P10	12		10	20
14L8 14P8	14		8	20
16L6 16P6	16		6	20
18L4 18P4	18		4	20
20L2 20P2	20		2	20
20C1	20		2 (Q & $\overline{Q}$)	20
Intermediate PAL:				
20L10	12	8	2	40*
Medium complexity PALs:				
16L8 16H8 16P8		6	2	64*
20L8	14	6	2	80*
20S10	12	8	2	80*

* – includes AND gates for tri-state control

The only PAL which does not fit into the groups exactly is the 20L10. This device features bidirectional I/O but has only three AND gates per output, so it falls between the low-complexity and medium-complexity categories.

The second table in this section covers the performance options available for each of the architectures listed in the first table. These are categorised by complexity and package size; a null entry in any cell means that this combination is not available at the time of writing. Because power consumption of CMOS varies with frequency it is quoted at d.c. and 10 MHz frequencies. The figures quoted for each technology are supply current(mA)/delay time (ns):

Device			*Bipolar technology*			
Structure	*STD*	*STD-2*	*‘A’*	*‘A – 2’*	*‘A – 4*	*‘B’*
LC20	90/35	45/60	90/25			
LC24	100/40					
20L10	165/50		165/30			
MC20	180/35		180/25	90/35	50/55	180/15
MC24	240/40		210/25	105/35		
			CMOS Technology			
	CMOS1		*CMOS2*	*CMOS3*		
LC20				20/50/55		
MC20	0/50/125		20/60/35	50/80/55		

From the above table it may be seen that there is much more choice in bipolar technology than CMOS. This is due, in part, to the fact that bipolar PALs have been established for a lot longer than CMOS, which have tended to be introduced with more complex structures, as will be discussed in Chapter 5. Even the CMOS3 family includes the Xpandor cell described earlier and is therefore not a truly fair comparison.

As an example of the actual PAL structures available, Figure 4.19 shows the PAL12L6 and Figure 4.20 the PAL20L8.

4.3.4 Structures for Address Decoding

4.3.4.1 Application Requirements

We have already looked at an example in which the PLE5P8 was used to decode the addresses from a microprocessor, in order to select ROM, RAM and I/O. In

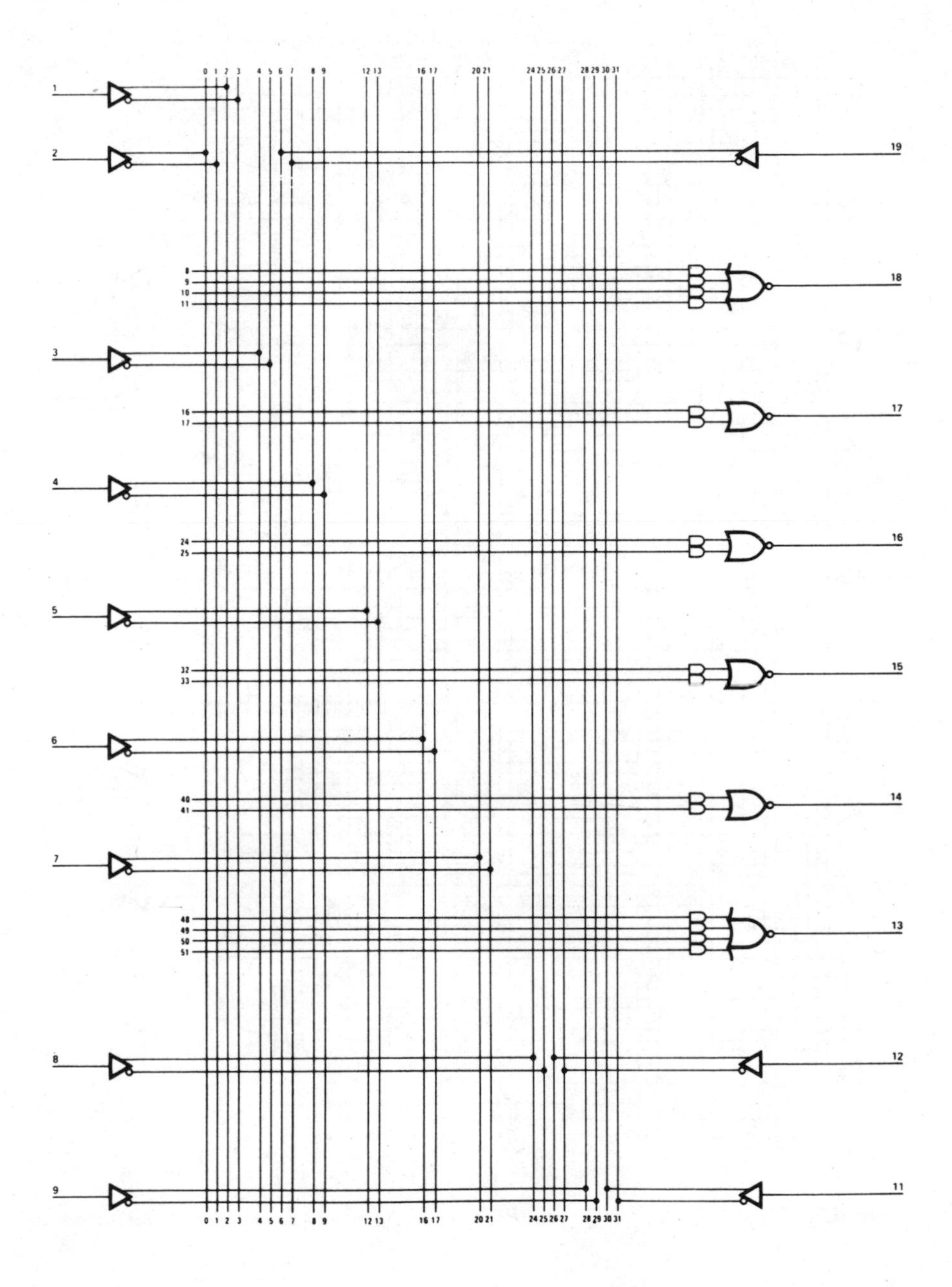

Fig. 4.19 PAL12L6 circuit diagram. *(Reproduced by permission of Monolithic Memories Inc.)*

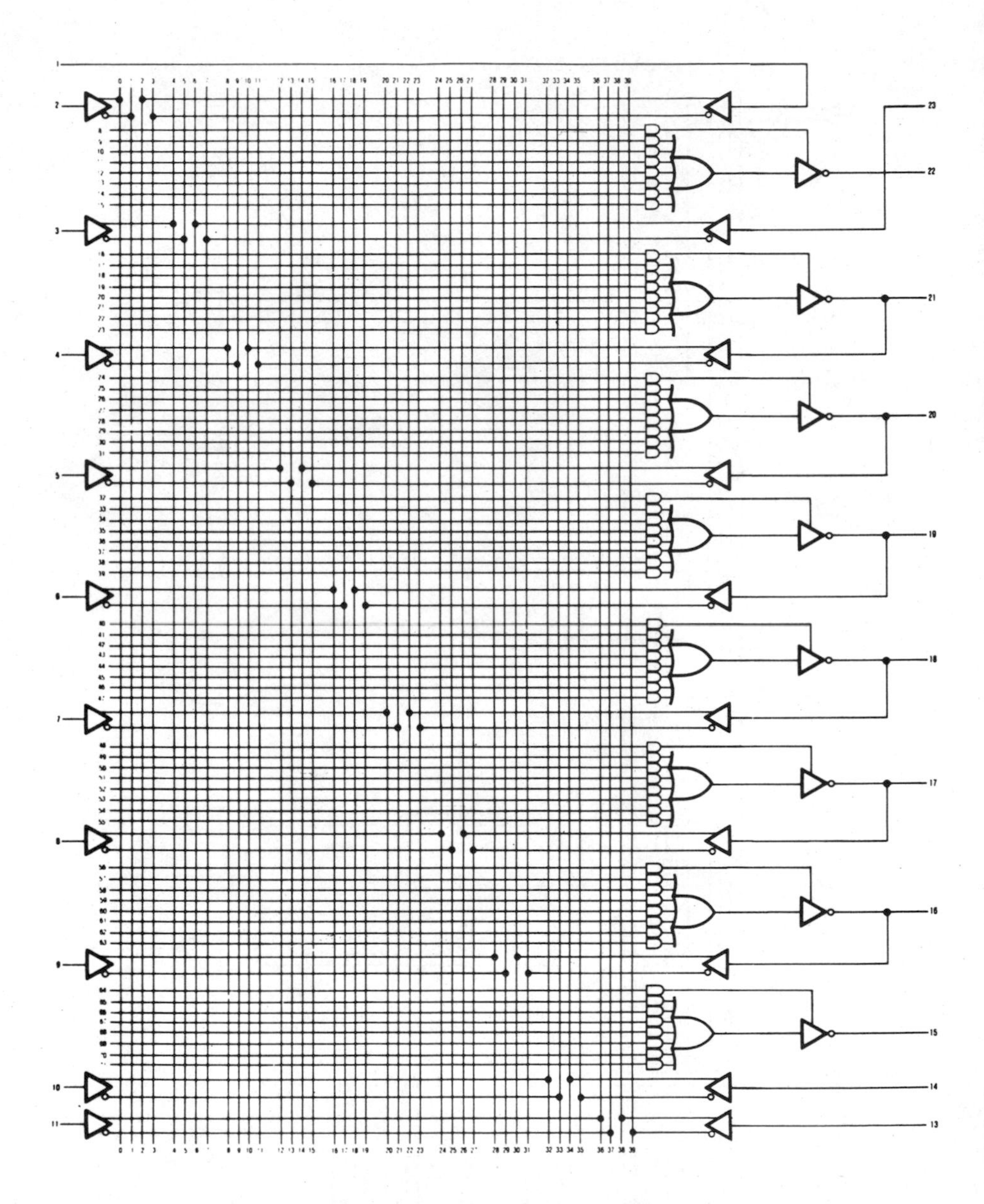

Fig. 4.20 PAL20L8 circuit diagram. *(Reproduced by permission of Monolithic Memories Inc.)*

many systems it is possible to define the address range for each selected block by a single AND gate. The condition for this is that the address range covered is a power of 2 and, when the lower limit is divided by the address range, the result is an even number. If an address decoder is being designed for such a system then

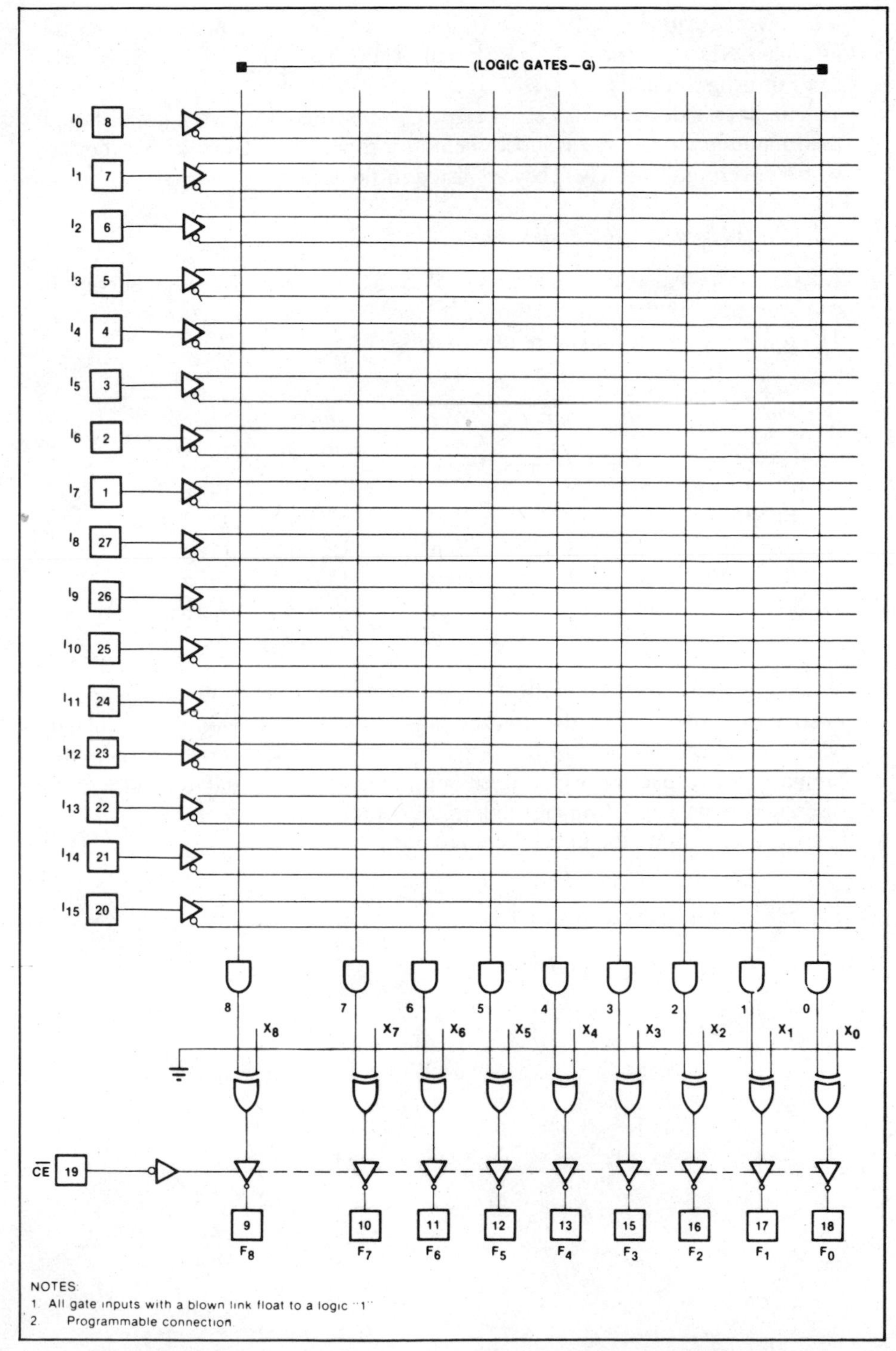

Fig. 4.21 PLS103 circuit diagram. *(Reproduced by permission of Mullard Ltd)*

it follows that either a PLA or a PAL is more complex than necessary. Because just one AND gate is needed for each output, there is no need for an OR matrix, fixed or programmable.

A number of devices have been designed for just this eventuality and, although their manufacturers have included them in a range of PALs or PLAs, they are worth covering separately. They are listed in the following section.

4.3.4.2 Address decoder availability

All the address decoders, except the PLS151, have fixed input and output configuration; all the PLS151 outputs are configured as bidirectional I/O pins. The following devices are currently available:

Part number	*Input pins*	*Output pins*	*Prop. delay*	*Supply current*	*Output structure*
PLS103	16	9	35 ns	170 mA	Prog. TS
PLS151	6	12	25 ns	155 mA	Prog. B I/O
PLS162	16	5	30 ns	155 mA	Prog. TS
PLS163	12	9	30 ns	155 mA	Prog. TS
PAL6L16	6	16	25 ns	90 mA	Active LOW
PAL8L14	8	14	25 ns	90 mA	Active LOW

While the above are specifically designed for address decoding they may find other uses. Some of them do offer the possibility of a sixteen-input AND gate which is far wider than can be found in any standard discrete logic family. Similarly, a gate package with sixteen outputs has the potential to replace many discrete gate packages from a standard logic family.

Figure 4.21 shows the full circuit diagram of the PLS103.

Chapter 5

Sequential PLDs

5.1 REGISTERED PLDs

5.1.1 PROM structures

5.1.1.1 Synchronisation

The simplest way to derive a sequential PLD from combinational devices is to feed the outputs into a D-type register. This structure is illustrated in Figure 5.1. The effect of this is to synchronise and store the outputs when the register receives an active clock edge. In every other respect the device acts in the same way as its combinational parent. A number of PROMs or PLEs are available with registered outputs, and this enhancement opens the door to more complex applications than are possible with the basic unregistered parts.

A common technique which is made possible by synchronous outputs is *pipelined operation*. This is a method of reducing the effective delay in systems where a number of operations are carried out in successive stages. An example of a simple pipelined system is the microprocessor instruction cycle, described in Section 3.3.2.1, although this uses a D-latch to store the address. The principle remains the same, however; the processor sends the address to the program memory at one stage of the cycle even though it does not require the result back until later in the cycle. Pipelining is often used in arithmetic operations to speed up throughput; Figure 5.2 shows a system for implementing Pythagoras' Theorem with PLEs.

Two methods are illustrated: the first uses combinational PLEs and, assuming a delay through each PLE of 30 ns, a result can be obtained every 90 ns when all the PLEs have settled. If a registered PLE with set-up time of 30 ns, hold time of 0 ns and delay of 15 ns is used, a result can be obtained every 45 ns even though

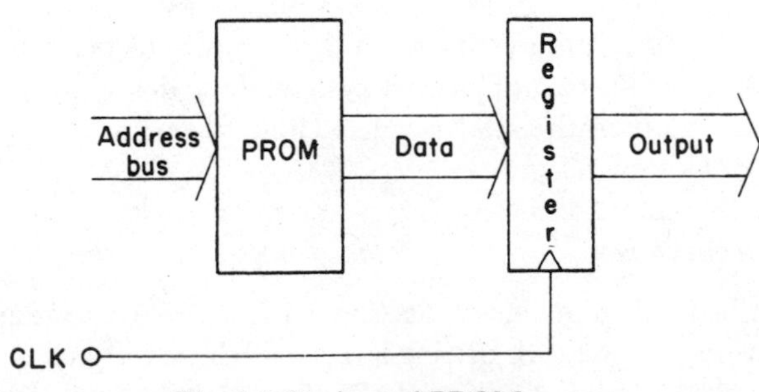

Fig. 5.1 Registered PROM.

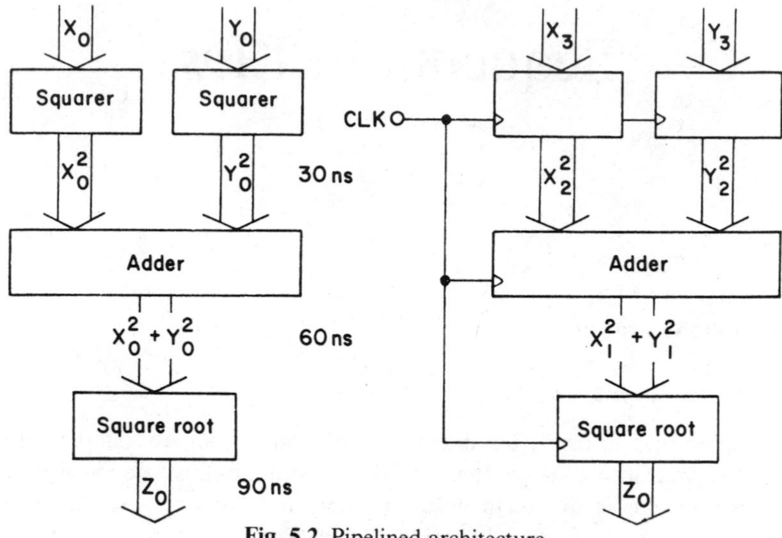

Fig. 5.2 Pipelined architecture.

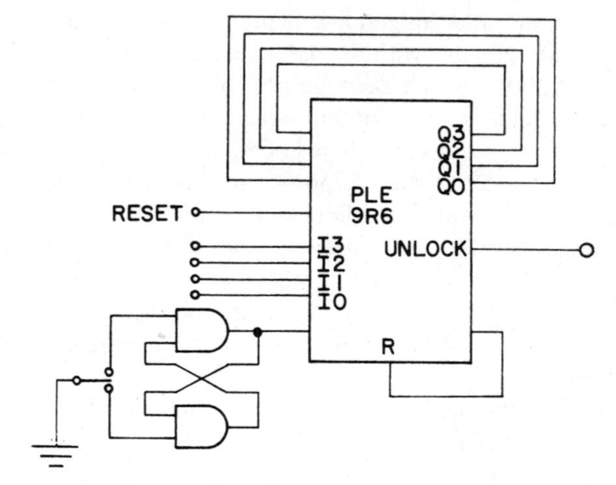

Fig. 5.3 Extended combination lock built in a registered PROM.

the data takes 135 ns to progress through the system. The benefit is obtained if a pipelined circuit forms part of a larger system; if the delays are comparable to those shown above then the system will be able to operate twice as fast as if unregistered PLEs were used.

5.1.1.2 State machines

We first described state machines in Section 3.4.2, where we saw that they can be constructed from a logic block driving a register whose outputs are fed back as further inputs to the logic. A registered PLE can thus be adapted to be a state machine by feeding some or all of the outputs back externally. The extended combination lock, which we used as an example in Section 3.4.3, fits well into a registered PLE as can be seen in Figure 5.3. The device required would be a PLE9R6; as with combinational devices the '9' and '6' refer to the number of inputs and outputs respectively, while the 'R' denotes the fact that the PLE contains registers.

The logic equations for the device remain the same as in Section 3.4.3.3. These need to be compiled into a format suitable for entry into a PLE, which is basically a memory device. To do this by hand is very tedious; in this example there are 512 possible input combinations and each has to be analysed to determine the output state for that combination. It may help if a Karnaugh map can be drawn for each output but these become unwieldy with more than about eight inputs. In practice, a computer may be used to generate the data; details of available programs will be given in a later chapter.

5.1.1.3 Diagnostic PROMs

We will cover the problems and requirements of testing in a later chapter, but there is a particular structure which makes systems more testable. This is called *Level Sensitive Scan Design*, usually shortened to LSSD. It is intended to solve the problems of observability and testability in complex systems like, for example, the pipelined Pythagoras calculator. If an incorrect answer is obtained from this system there is no way of knowing in which section the fault lies, without looking at the intermediate results. LSSD allows systems to be put into a test mode where the inputs to individual sections can be forced to a known state, and the resulting outputs observed. The way in which this can be done is shown in Figure 5.4.

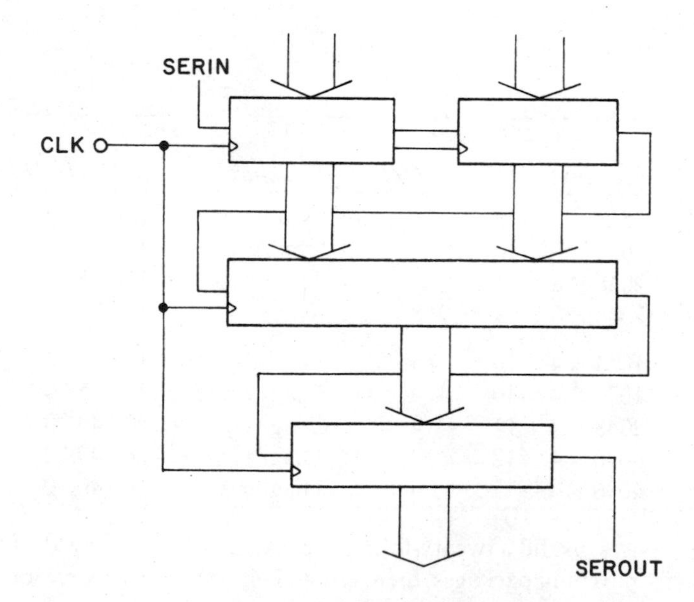

Fig. 5.4 LSSD system architecture.

In synchronous systems the output registers are given the ability to be loaded serially from a single external test point, by wiring all the registers into one long shift register. In test mode the data is fed in by clocking for as many cycles as there are cells in the registers. The system is then allowed to operate for a fixed number of cycles and the result clocked out serially. By comparing the result with the expected data the system can be tested and faults diagnosed with some precision. A family of *Diagnostic PROMs* has been developed specifically for this type of application.

Diagnostic PROMs, illustrated in Figure 5.5, include a second register called the *shadow register*. This can be loaded and emptied serially, and the contents transferred to and from the output register when the PROM is in test mode. These extra facilities use up four pins so the test facility is built-in with very little overhead on device complexity.

5.1.1.4 Registered PROM availability

Registered PROMs are manufactured in both bipolar and CMOS technology; they have similar speed characteristics, but CMOS devices consume about half the power. The output enable may be applied in one of two ways, asynchronously when it acts in the same way as in an unregistered PROM, or synchronously when it is loaded into a flip-flop by the clock. Smaller registered PROMs include both options but the larger PROMs have only one or the other as there are not enough pins to allow both inputs. The asynchronous parts have the designation 'RA' whilst the synchronous parts are given 'RS'. The diagnostic PROMs are generally known by a manufacturer's part number rather than a PLE type number; the table of available devices is given below, with delays in nanoseconds:

Part Number	*Memory*	*I/Ps*	*O/Ps*	*Structure*	*Del/S-U/Hold*
PLE9R8	512 × 8	9	8	Syn/Asyn	15/ 30/ 0
PLE10R8	1024 × 8	10	8	Syn/Asyn	15/ 30/ 0
PLE11RA8	2048 × 8	11	8	Async	15/ 35/ 0
PLE11RS8	2048 × 8	11	8	Sync	15/ 35/ 0
63DA441	1024 × 4	10	4	Diag/asyn	18/ 35/ 0
63DA442	1024 × 4	10	4	Diag/syn/as	18/ 35/ 0
63DA841	2048 × 4	11	4	Diag/asyn	20/ 40/ 0
63D1641	4096 × 4	12	4	Diag/asyn	20/ 40/ 0
63DA1643	4096 × 4	12	4	Diag/init	20/ 40/ 0

The largest devices just fill a twenty-four-pin package so, unless new devices are put into higher pin count packages, there is not likely to be much extension to the above range.

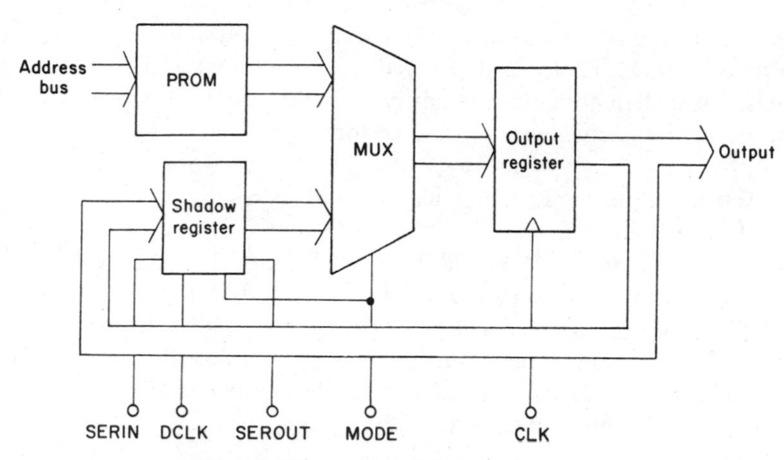

Fig. 5.5 Diagnostic PROM structure.

5.2 REGISTERED PALs

5.2.1 Simple structures

5.2.1.1 PLE Restrictions

As in combinational logic, we find that registered PLEs are limited by the number of inputs which they can support. In applications such as pipelining this is not a disadvantage, because the purpose of the output register is to lock out the result of one stage of logic until the following stage is ready to accept it. The restriction on numbers of inputs can be a problem but no more than in combinational systems; after all, a pipelined system is no more than a series of combinational logic blocks buffered from each other. Indeed, a very common application is in arithmetic blocks where many AND terms are often needed and using a PAL or PLA is likely to be even more restricting.

Problems are far more likely to occur in state machine applications. In these cases some, or all, of the outputs must be fed back to the input so that the present state of the system can be incorporated into the state jump decision. In the worst case of a system needing eight outputs fed back, the largest registered PLE can accommodate only three more inputs. Furthermore, to generate the logic from even a simple state table is a very tedious task without the use of a computer, as we saw in the example which we discussed. In the same way that we derived the PLA and PAL structures from PROMs, we will see how registered PLAs and PALs are simpler to use and offer more scope for integrating state machines.

5.2.1.2 PALs with output registers

The simplest enhancement to a combinational PAL is the addition of an output register, as in Figure 5.6. In this class of device each output is loaded into a D-type flip-flop on an active clock edge, the registered output being fed back internally to the AND gate array. This structure is similar in principle to the registered PROM, except that the feedback is internal and does not use up inputs. The benefit of this has already been mentioned; in addition, the design is much simpler as many fewer AND gates need defining.

In practice, a family of registered PALs is derived from one combinational PAL. For example, the PAL16L8 with registers becomes the PAL16R8, the 'R' indicating a registered part, and there are two other PALs with a similar structure; these are the PAL16R4 and PAL16R6. The PAL16R4 has four registered outputs and four bidirectional pins, while the PAL16R6 has six registered and two bidirectional pins. This allows state machines with up to twelve inputs to be built in the smallest package, because bidirectional pins can function as inputs. Alternatively, it is possible to mix combinational logic and sequential logic within one device.

5.2.1.3 Design example

We can see how a registered PAL can be used for the enhanced combination lock, described in Section 3.4.3. One problem with the equations of Section 3.4.3.3 is that they include terms such as:

$$\overline{(I3 * \overline{I2} * \overline{I1} * \overline{I0})}$$

This could be expanded by DeMorgan's laws into four OR gates but then Q3 and Q2 would need twelve AND gates each. Simple PALs have only eight AND gates

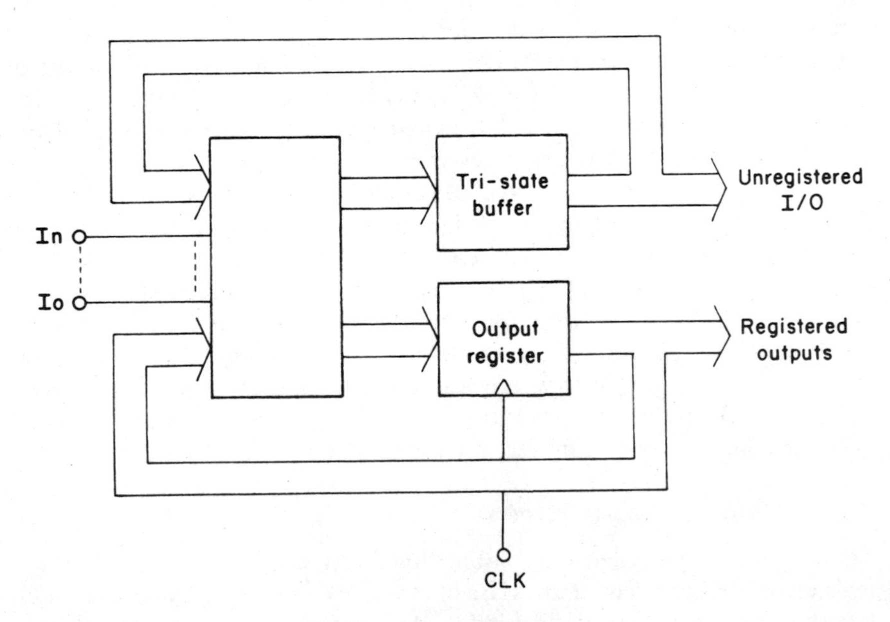

Fig. 5.6 Registered PAL.

per output so an alternative must be sought. If a PAL16R4 is used, then the bidirectional pins can be used to form the complement functions, such as that above. Calling these intermediate functions C3, C2 and C1 respectively, the equations become:

$$\overline{C3} = I3 * \overline{I2} * \overline{I1} * \overline{I0}$$
$$\overline{C2} = \overline{I3} * \overline{I2} * \overline{I1} * \overline{I0}$$
$$\overline{C1} = I3 * \overline{I2} * \overline{I1} * I0$$

$$\begin{aligned} Q3 := {} & Q2 * \overline{Q1} * \overline{Q0} * C3 \\ & + Q2 * \overline{Q1} * Q0 * C2 \\ & + Q2 * Q1 * \overline{Q0} * C1 \end{aligned}$$

$$\begin{aligned} Q2 := {} & Q2 \\ & + \overline{Q2} * \overline{Q1} * \overline{Q0} * C3 \\ & + \overline{Q2} * \overline{Q1} * Q0 * C2 \\ & + \overline{Q2} * Q1 * \overline{Q0} * C2 \end{aligned}$$

$$\begin{aligned} Q1 := {} & \overline{Q3} * \overline{Q1} * Q0 * \overline{C2} \\ & + \overline{Q3} * Q1 * \overline{Q0} * \overline{C1} \end{aligned}$$

$$\begin{aligned} Q0 := {} & \overline{Q3} * \overline{Q1} * \overline{Q0} * \overline{C3} \\ & + \overline{Q3} * \overline{Q2} * \overline{Q1} * Q0 * C2 \\ & + \overline{Q3} * Q1 * \overline{Q0} * \overline{C1} \end{aligned}$$

$$UNLOCK = Q1 * Q0$$

As the last equation shows, we can use the fourth bidirectional pin as an output to provide the UNLOCK signal. Care must be taken with the output polarity; the PAL16R4 has active LOW outputs, so the actual signals will be the complement of the above. A PAL16RP4 is available with programmable output polarity to overcome this problem. Also we have not been able to include the asynchronous reset, as this feature is not included in the simple PALs.

5.2.1.4 Hold and toggle methods

Registered PALs use D-type flip-flops almost exclusively. Referring back to our description of flip-flop types in Section 3.3.2.4 it is apparent that D-type flip-flops do not possess the inbuilt functions 'hold' and 'toggle'. These functions need to be incorporated in many sequential logic functions, such as counters and state machines in general. In order to incorporate these functions there are two possible approaches.

One way, which we followed in designing the enhanced combination lock, is to analyse the individual bits to define all the conditions which result in a HIGH being loaded. This means that each output is treated as a combinational logic circuit with the flip-flop appearing as an appendage. If we are to get the most out of registered PALs it is better to understand how the various sequential functions can be designed directly into the PAL. Loading a HIGH or LOW directly is straightforward as these functions exist implicitly in the D-type truth table; let us therefore turn our attention to the hold and toggle.

Hold, or don't change, means that if the output is already HIGH a HIGH must be loaded by the next clock edge. If the output is LOW then it must remain LOW, which it will do if none of the AND gates feeding the flip-flop is true. This result can be achieved by feeding the output back to the input and gating it with the condition for holding. In other words an equation can be written as:

$$Q = \text{HOLD CONDITION} * Q$$

This has the desired effect because the equation is only true if Q is HIGH when the hold condition is present.

The requirement for toggling is just the opposite; a HIGH must become LOW after the clock while a LOW must change to HIGH. If we feed back the complement of the output then the output will toggle. The equation for this is:

$$Q = \text{TOGGLE CONDITION} * \overline{Q}$$

In this case, if the toggle condition is present, the equation is only true if the output is LOW.

5.2.1.5 Building a counter

We can apply these techniques to seeing how easily a counter may be defined as logic equations for a registered PAL. We can examine a four-bit counter bit by bit and build the equations from the functions we require, instead of by constructing the state diagram. In addition to the basic counting function let us include a parallel load controlled by a 'LOAD' input, which enables data on four inputs to be put into the respective bits of the counter, and disables the count function while it is present.

The least significant bit (LSB) of the counter, Q0, has to toggle for every count and also has to be set HIGH when I0 is HIGH in the load condition. Thus the equation for Q0 is:

$$\begin{aligned} Q0 :=\ & \overline{\text{LOAD}} * \overline{Q0} \text{ (toggle)} \\ & + \text{LOAD} * I0 \text{ (load)} \end{aligned}$$

The next bit, Q1, toggles when Q0 is HIGH, holds when Q0 is LOW and loads when I1 is HIGH so the equations are:

$$\begin{aligned} Q1 :=\ & \overline{\text{LOAD}} * \overline{Q1} * Q0 \text{ (toggle)} \\ & + \overline{\text{LOAD}} * Q1 * \overline{Q0} \text{ (hold)} \\ & + \text{LOAD} * I1 \text{ (load)} \end{aligned}$$

The third bit toggles when both Q1 and Q0 are HIGH but must hold under all other count conditions; in other words, when either Q1 or Q0 are LOW. Q2 is loaded from I2 giving the following equations:

$$\begin{aligned} Q2 :=\ & \overline{\text{LOAD}} * \overline{Q2} * Q1 * Q0 \text{ (toggle)} \\ & + \overline{\text{LOAD}} * Q2 * \overline{Q1} \text{ (hold)} \\ & + \overline{\text{LOAD}} * Q2 * \overline{Q0} \text{ (hold)} \\ & + \text{LOAD} * I2 \text{ (load)} \end{aligned}$$

The equations for the *Most Significant Bit* (MSB) can be written down by following a similar argument:

$$\begin{aligned} Q3 := &\ \overline{LOAD} * \overline{Q3} * Q2 * Q1 * Q0 \text{ (toggle)} \\ &+ \overline{LOAD} * Q3 * \overline{Q2} \text{ (hold)} \\ &+ \overline{LOAD} * Q3 * \overline{Q1} \text{ (hold)} \\ &+ \overline{LOAD} * Q3 * \overline{Q0} \text{ (hold)} \\ &+ LOAD * I3 \text{ (load)} \end{aligned}$$

Similar reasoning will allow equations to be written directly for any system which can be described in terms of the standard sequential functions.

5.2.2 Enhanced registered PAL structures

5.2.2.1 Limitations of simple PALs

The examples described so far have, not unnaturally, fitted into simple registered PALs. There are, nevertheless, some factors which limit the complexity of circuits which can be implemented in the structures described above. Perhaps the most obvious of these is the number of AND gates available for each output. The simple PALs have, as a rule, just eight gates per output which does limit the complexity of logic which the PAL can support.

Another although, arguably, less demanding restriction is the balance between inputs, registered outputs and unregistered outputs. Simple PALs overcome this, to a limited extent, by offering a range with differing configurations. The divisions are somewhat artificial and it would be useful to have more choice of pin function as the circuits become more complex.

Finally, a feature which is vitally important when it comes to testing, as we shall see later, but may also play an important role in normal operation, is the ability to set the output register to a known state independent of the logic inputs. In showing how the enhanced combination lock can be fitted into a PAL16R4 we saw that we could not include the reset function. This would have to be implemented externally if we were designing an actual circuit but would clearly give advantages if it were included internally. The chief benefit is that it would enable the circuit to be set to a known state; as it stands there is no way of knowing which state the circuit will be in, even when it is first switched on.

One way round this is to incorporate an external power-on using a spare input to control the setting-up of the register. Figure 5.7 shows a typical configuration; the capacitor is charged slowly through the resistor so 'PON' stays LOW for some time after the power supply has stabilised. The PAL equations have to be modified as follows:

$$\begin{aligned} Q := &\ \overline{PON} \\ &+ PON * \text{NORMAL FUNCTION} \end{aligned}$$

The PAL is thus set with all its flip-flops HIGH until PON goes HIGH when normal operation can begin. It is preferable for this function to be included internally to the PAL.

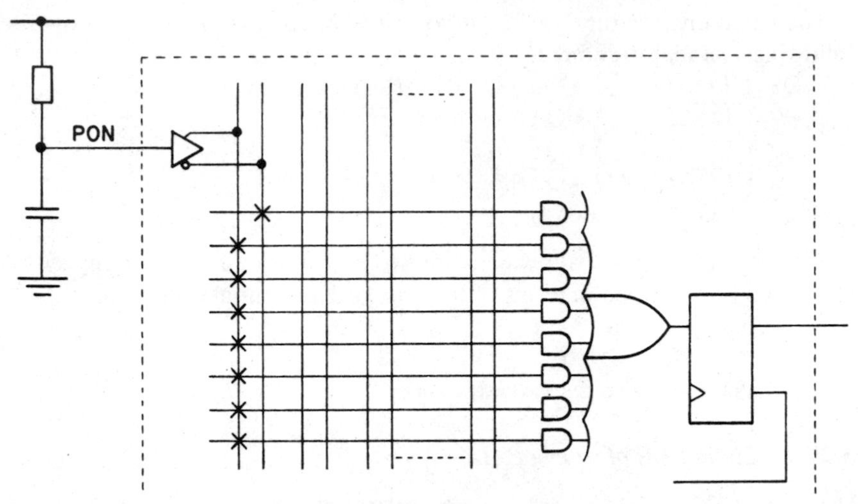

Fig. 5.7 External power-up circuit.

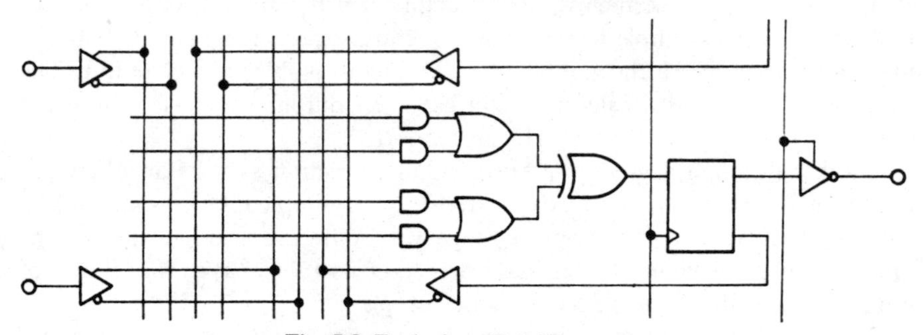

Fig. 5.8 Exclusive-OR PAL structure.

		LD	0	0	0	0	1	1	1	1
		I3	0	0	1	1	1	1	0	0
		Q3	0	1	1	0	0	1	1	0
Q2	Q1	Q0								
0	0	0		H	H		H	H		
0	0	1		H	H		H	H		
0	1	1		H	H		H	H		
0	1	0		H	H		H	H		
1	1	0		H	H		H	H		
1	1	1	H			H	H	H		
1	0	1		H	H		H	H		
1	0	0		H	H		H	H		

Fig. 5.9 Karnaugh map – divide-by-16 counter bit 'Q3'.

5.2.2.2 Exclusive-OR PALs

In Section 4.3.1.2 we saw that some benefits can be obtained by putting exclusive-OR gates immediately before the outputs. A range of registered PALs with internal exclusive-OR gates has been developed with the structure illustrated in Figure 5.8. This structure is particularly useful in applications involving arithmetic and counting, as we may see by looking back at the counter described in the previous section. Inspection of the logic equations shows that their number increases by one for each bit of the counter. Eventually, for a larger number of bits or a more complex counter, there would not be enough AND gates to build the counter. The Karnaugh map for Q3, which is shown in Figure 5.9 just for the counter bits, indicates how an exclusive-OR PAL can overcome this problem. The equation for Q3 can be written:

```
Q3 := Q3
   :+: Q2 * Q1 * Q0
```

In this format the Q3 term forms the hold function, the other term is the toggle function. Logically it works because one side of the exclusive-OR is HIGH and the other side LOW when Q3 is HIGH, except when Q2, Q1 and Q0 are HIGH. Similarly, both sides are LOW when Q3 is LOW unless the other three bits are HIGH. Q3 is taken HIGH when both sides are different but goes LOW when they are the same. The parallel load function is not affected because the feedback bits are gated with $\overline{\text{LOAD}}$ in count mode so all the count terms will be LOW when LOAD is HIGH. The load function can be combined into either side of the exclusive-OR gate because all other terms are LOW when loading. The full equation for Q3 is thus:

$$
\begin{aligned}
Q3 &:= \overline{\text{LOAD}} * Q3 \\
&\quad + \text{LOAD} * I3 \\
&:+: \overline{\text{LOAD}} * Q2 * Q1 * Q0
\end{aligned}
$$

Two of the series of exclusive-OR PALs have additional logic circuitry to make arithmetic applications easier to implement. Both the PAL16X4 and PAL16A4 have the network shown in Figure 5.10 on four inputs. This network includes the four logic functions $I + Q$, $I + \overline{Q}$, $\overline{I} + Q$ and $\overline{I} + \overline{Q}$. By connecting one or more of these functions into the AND array virtually any logical combination of input and output can be used as an input to the AND array.

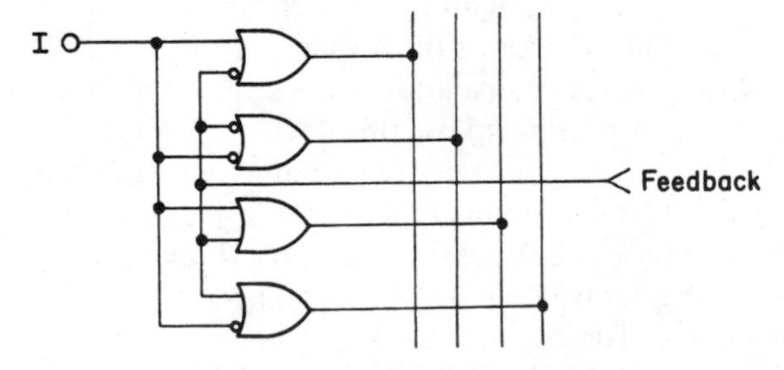

Fig. 5.10 'Arithmetic' PAL inputs.

The PAL16A4 has additional AND gates pre-programmed into the logic array to provide the carry functions for arithmetic circuits. This device is very specialised in its application and therefore does not fit so well into the concept of programmable logic. A detailed application note is available from the manufacturer covering its use in arithmetic circuits.

The 'X' designation is used to indicate exclusive-OR PALs; 'A' is used for the arithmetic PAL.

5.2.2.3 'Versatile' PALs

Many of the shortcomings of the simple registered PALs have been improved by the so-called Versatile PAL. This PAL, part number 22V10, has been designed with a pinning and architecture which enables it to act like any other simple twenty-four-pin registered PAL. At the same time, extra features have been included which enable it to fit applications where simple PALs cannot be used.

As its part number implies it has ten output pins but they do not all have the same number of AND gates feeding them. The least powerful have eight gates per output, as the simple PALs, the most powerful have sixteen gates and there are outputs with intermediate numbers. It is possible to tailor the output selected to the number of AND gates needed to implement the function required from that output. For example, building a large counter does not present a problem because the bits requiring the most terms to define the hold function can occupy the high term outputs.

Perhaps the most significant innovation in the 22V10 is the output macrocell structure. This is illustrated in Figure 5.11 where it may be seen that there are two extra fuses associated with each output. These fuses control a four-input multiplexer which drives the output buffer, and a two-input multiplexer which feeds back to the AND array. The multiplexer inputs are the direct output from the logic array and its complement, and the registered output and its complement. Feedback can be taken direct or registered. In addition there is a controlled tri-state buffer between the multiplexer and the direct feedback connection so the output can be switched off and the pin used as an input. Each of the ten output pins can therefore be an input, an active HIGH or active LOW output, or a registered active HIGH or active LOW output with a tri-state control option for each of the output modes.

There are also two extra AND gates in the logic array; these provide a synchronous preset and an asynchronous reset common to all the flip-flops in the output register. This overcomes the problem, discussed earlier, of not being able to start the state machine in a known state for testing or clean operation.

Finally, because it is possible to use this device as either a combinational or sequential PAL, the pin used for the clock input is also connected to the AND array as an input. A block diagram of this Versatile PAL is shown in Figure 5.12 to provide a summary of all the features described above, and to indicate the allocation of AND gates between the various output pins.

The output macrocell described above is not unique to the 22V10, although no other PALs have quite the same distribution of AND gates among the outputs.

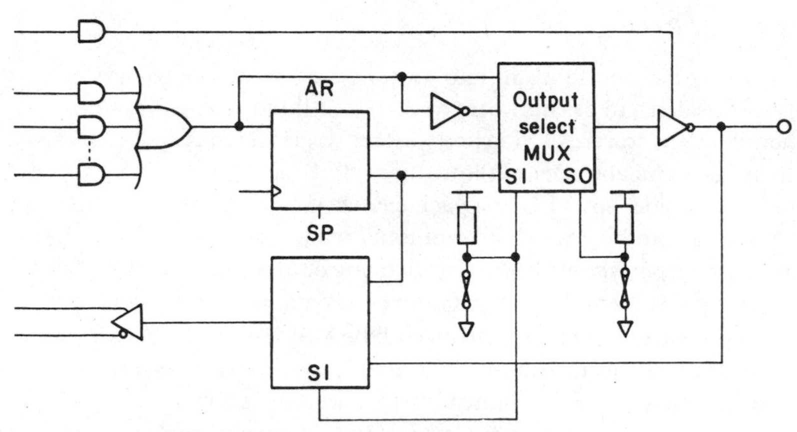

Fig. 5.11 'Versatile' PAL output macrocell.

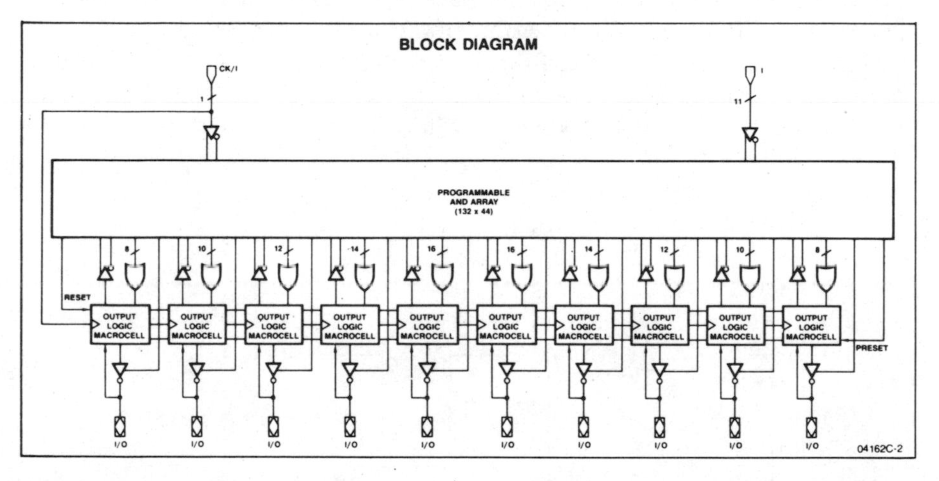

Fig. 5.12 'Versatile' PAL block diagram. *(Reproduced by permission of Advanced Micro Devices)*

The two smallest members of the EP-series of PAL use the twenty- and twenty-four-pin packages, and both feature a similar macrocell or register bypass output types.

Another recent introduction is *Generic Array Logic*, or GAL (a trademark of Lattice Semiconductor Corp), which can replace any simple PAL by programming its architecture bits appropriately. This device, which is only available in electrically erasable CMOS, has an output macrocell similar to the 22V10 or EP300 but with extra features to allow true PAL replaceability. The macrocells can also be programmed individually to give the same versatility as the 22V10 and EP300 in terms of mixing output types.

5.2.2.4 LSI PALs

The devices described so far fit into a twenty- or twenty-four-pin package and may be considered to be equivalent to discrete MSI logic circuits. However, just as there is a trend towards LSI in both discrete logic and microprocessors, so LSI PALs are also available. One of the features of LSI is an increased number of I/O pins so we consider any PLD in a package size greater than twenty-eight pins to be LSI. An assumption which is commonly made when specifying PALs is that at least half the pins should be capable of being outputs so, unless the logic power of each output is decreased, a higher pin count implies a large number of gates.

One approach to LSI has been to extend the MSI-registered PALs by doubling the number of inputs and outputs and including some of the features already described to make the devices more versatile. Two devices in this category are the 32R16 and 64R32, which is a doubled 32R16. The input configuration for these devices is similar to any standard PAL with, in the case of the 32R16, sixteen dedicated inputs and sixteen outputs fed back to the AND array. The outputs are controlled in groups of eight rather than individually, as in the case of 22V10, each group of eight being formed from four doublets. The structure of an output doublet is shown in Figure 5.13.

Each doublet has sixteen AND gates which can be shared between the two

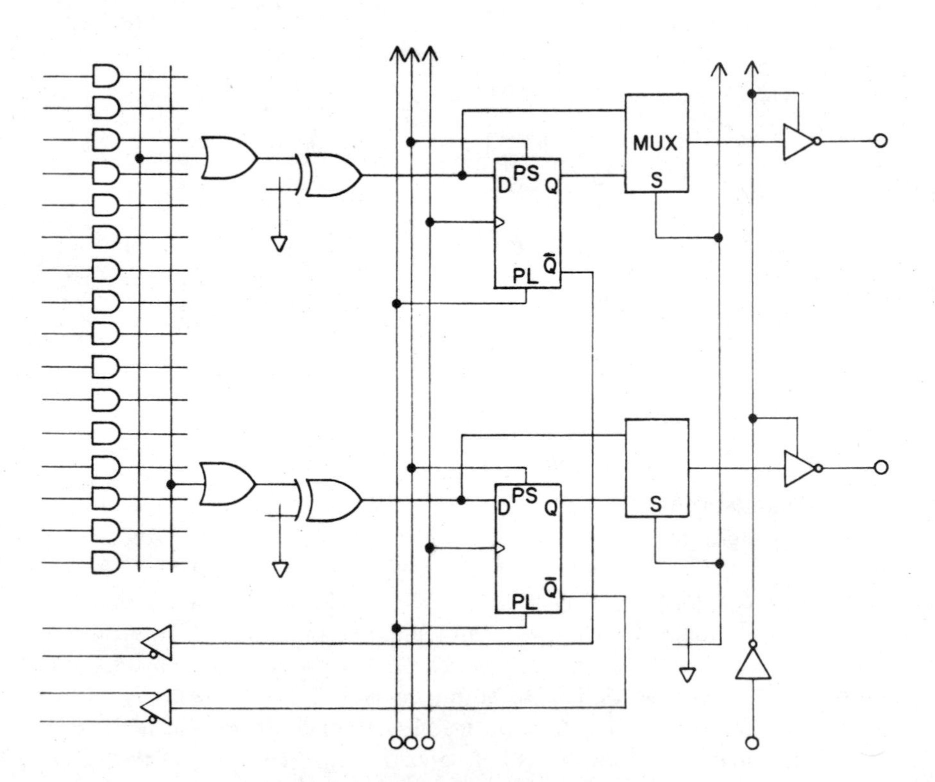

Fig. 5.13 PAL32R16 output cell.

outputs, which can be set active HIGH or active LOW independently. The output register can be bypassed to form a combinational output, but this facility must be applied to all outputs in the group of eight. Each group of eight shares a common clock, a common preset and a common tri-state enable; there is also a common 'preload'. This feature allows data present on the outputs, after they have been tri-stated, to be loaded asynchronously into the output register. Together with the preset, this makes initialisation for testing or start-up quite straightforward.

One feature which has been omitted is the bidirectional function. The PAL32R16 has sixteen inputs and sixteen outputs with no variation possible. Although the output groups are far more flexible than an individual PAL16R8, for example, the majority of applications are likely to be those which could be performed equally well by two MSI PALs, in the case of the PAL32R16, or four MSI PALs in the case of the PAL64R32. The advantage obtained by combining the PALs into one package is that each output can be formed from all the inputs and fedback outputs. One area where this could be useful is in interfacing to sixteen- and thirty-two-bit microprocessors if synchronous logic is preferred.

The other PAL family which has been extended into LSI is the EP series. We have already mentioned that the two smallest members of the family have an architecture which is similar to the 22V10, inasmuch as the 'output' pins can be

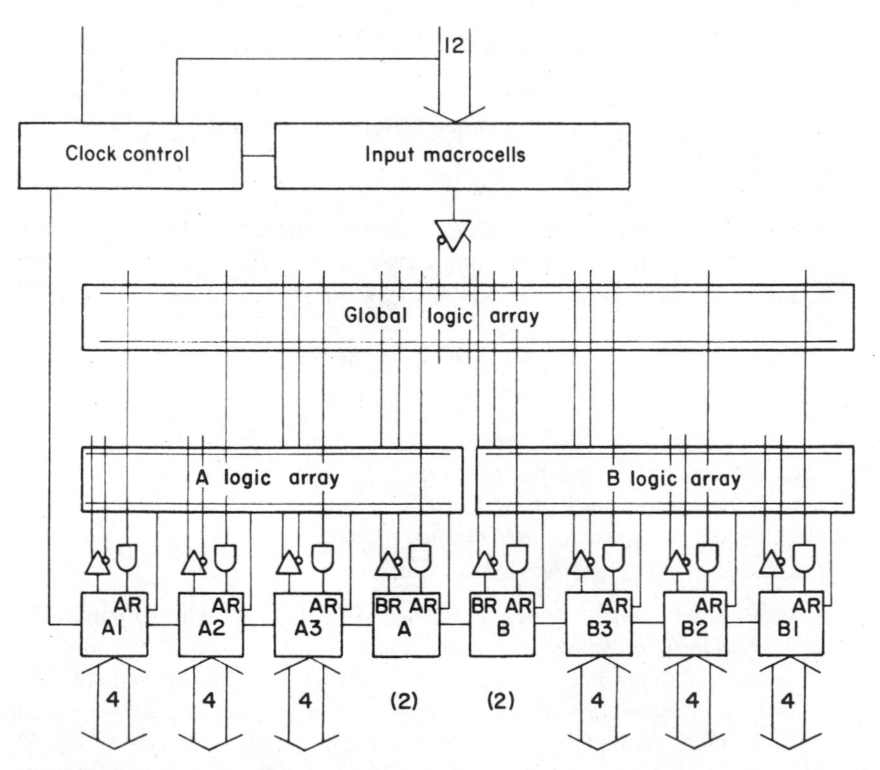

Fig. 5.14 EP1200 block diagram.

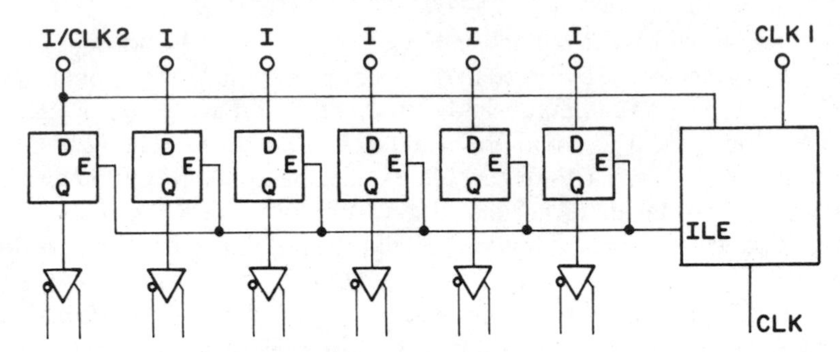

Fig. 5.15 EP1200 input block.

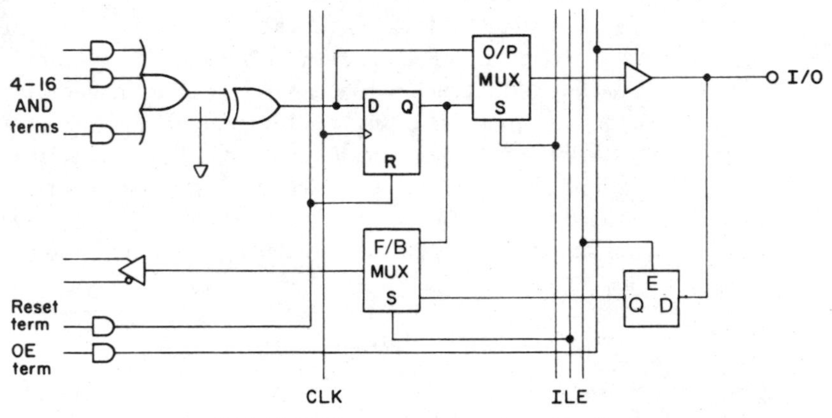

Fig. 5.16 EP1200 output register (one flip-flop).

configured in one of five different ways. The larger members of the family also feature this architecture, with some enhancements. We will examine one of these, the EP1200, in some detail in order to show how one trend in programmable LSI is progressing.

The block diagram is shown in Figure 5.14. The whole structure may split into functional blocks as follows:

- input block (twelve latchable inputs)
- output macrocells (six groups of four outputs)
- buried registers (two pairs of flip-flops)
- global bus (forty-eight-wide array input)
- local busses (two sixteen-wide array inputs)

The input block, which includes the clock control, is shown in Figure 5.15. There are two inputs to the clock control circuit, which has two outputs. One clock is used as a common clock to all the output registers, the second output is an optional latch enable which is connected to all the input latches. Three control bits define which sense of each input is used to provide the two outputs. All inputs are fed to the array via D-latches although two combinations of control bit cause the latches to be permanently transparent. The timing of the clocks is such that the inputs are latched for the half period before the active clock edge, if the same input is used for both. This will prevent any chance of metastability in the output register when the inputs are asynchronous.

There are twenty-four I/O pins grouped into six quartets. Each group has a common tri-state enable, reset, register bypass and feedback selection, as shown in Figure 5.16 for one flip-flop. Direct feedback from the pin also passes through a D-latch with the same control as the inputs, so each group may be dedicated as an input group. There is a variable number of AND gates connected into each output within the group and two of the groups also feature shared AND terms. The outputs from these two groups are fed back to the global bus while the outputs from the other four groups are fed back to their respective local busses.

In addition to the output registers there are two pairs of 'buried' registers. These are not connected to any output pins but are fed back to the global bus; they also have bypass and feedback selection and share a reset with their adjacent output register group. Their structure is similar to the output register without the output buffer and pin connections. A typical use for buried registers would be as an internal counter in order to implement an internal delay in a state machine application. They could also be used as we used the bidirectional pins in our PAL16R4 design for the enhanced combination lock.

The bus structure in the EP1200 was developed to make the addressing simpler for programming. As it stands, there are sixty-four inputs to each AND gate making them addressable by six bits. If all the inputs to the array were made available to the whole array then each AND gate would have eighty inputs, which would require an extra address line. The inputs and feedback from the buried registers and two of the output groups can be used as inputs to any AND gate – these are called the 'global bus'. Feedback from four of the output groups can be used as inputs to only half of the AND gates via the 'local bus'. This bus structure saves about 20 per cent of the programming requirement without significantly affecting the logic power of the EP1200, although it could place some constraint on pinning for some applications.

5.2.3 Registered PAL availability

5.2.3.1 Architecture and technology

As with other PALs, the devices described in this section are available in a number of technologies. The main series of PALs was developed in bipolar technology but the architectures have been copied in some CMOS variations. The three main CMOS varieties are u–v erasable cells, electrically erasable cells and polysilicon fuses, although some of the u–v erasable devices are made in opaque packages which are cheaper than the window packages which allow erasure. Like the unregistered parts, the specifications for the different CMOS registered parts are different for each manufacturer, so the technology table indicates the various options which are available.

In the table showing the available architectures, there are so many variations that the options used by each part are listed line by line. The PAL designation letters are:

R – 'plain' registered output (usually active-LOW)
RS – registered shared outputs
RP – registered programmable polarity outputs
V – 'versatile' outputs
X – exclusive-OR function internally
A – arithmetic function internally

5.2.3.2 Registered PAL summary

The first table shows the various architectures available:

Part Number	*Inputs*	*B I/O*	*Outputs*	*AND gates*	*Features*
Simple PALs:					
16R4 16RP4	8	4	4	64	o/p enable
16R6 16RP6	8	2	6	64	o/p enable
16R8 16RP8	8	–	8	64	o/p enable
20R4	12	4	4	64	o/p enable
20R6	12	2	6	64	o/p enable
20R8	12	–	8	64	o/p enable
Exclusive-OR PALs:					
16X4 16A4	8	4	4	64	o/p enable
20X4	10	6	4	40	o/p enable
20X8	10	2	8	40	o/p enable
20X10	10	–	10	40	o/p enable
AND sharing PALs:					
20RS4	10	6	4	80	o/p enable
20RS8	10	2	8	80	o/p enable
20RS10	10	–	10	80	o/p enable
Versatile PALs:					
16V8	8	8		74	set/reset
22V10	11	10		140	s/r v/trms
EP300 EP310	10	8		74	set/reset
EP600	4	16		160	s/r m/clk p/l

Part Number	*Inputs*	*B I/O*	*Outputs*	*AND gates*	*Features*
LSI PALs:					
32R16	16	–	16	128	
					shared gates, two clocks, preload
64R32	32	–	32	256	
					shared gates, four clocks, preload, set
EP1200	12	24	–	236	
					shared gates, variable terms, 4 buried regs latched inputs, reset, split bus

The performance parameters for each technology/architecture combination are:

Technology	*Delay*	*Clock*	*Set-up*	*Hold*	*Supply*
Bip 16R ser.	35 ns	16 MHz	35 ns	0 ns	180 mA
Bip 16R Aser.	25 ns	28 MHz	25 ns	0 ns	180 mA
Bip 16R A–2	35 ns	16 MHz	35 ns	0 ns	90 mA
Bip 16R A–4	55 ns	11 MHz	60 ns	0 ns	50 mA
CMOS ps 16R	125 ns	n.a.	n.a.	n.a.	5 mA/MHz
CMOS ep 16R	35 ns	16 MHz	20 ns	0 ns	5 mA/MHz
Bip 20R Aser.	25 ns	28 MHz	25 ns	0 ns	210 mA
Bip 16X/16A	35 ns	14 MHz	45 ns	0 ns	225 mA
Bip 20X ser.	50 ns	12 MHz	50 ns	0 ns	180 mA
Bip 20RS ser.	35 ns	20 MHz	35 ns	0 ns	240 mA
CMOS ee 16V8	25 ns	25 MHz	20 ns	0 ns	140 mA
Bip 22V10	25 ns	n 28 MHz	20 ns	0 ns	180 mA
CMOS EP300	35 ns	25 MHz	35 ns	0 ns	5mA/MHz
Bip 32R16	40 ns	16 MHz	40 ns	0 ns	280 mA
Bip 64R32	50 ns	16 MHz	40 ns	0 ns	640 mA
CMOS EP1200	50 ns	15 MHz	40 ns	0 ns	10 mA/1MHz

As with the combinational PALs, the performance of CMOS is better at low frequencies, in terms of power dissipation, but there is not much advantage at operating frequencies approaching the maximum. In general, it is better to select the architecture which is most suitable, irrespective of technology, and then select the technology which is most cost-effective for the application. The economics should take into account the cost of power supplies and cooling as well as the cost of the devices themselves. Given a straight choice between CMOS and bipolar technology, CMOS may well be cost-effective only at low frequencies and if the logic circuit is the major consumer of power within the system.

Figure 5.17 shows the full circuit diagram for a simple registered PAL, the PAL16R4, while a more complex registered PAL, the PAL22V10, is illustrated in Figure 5.18.

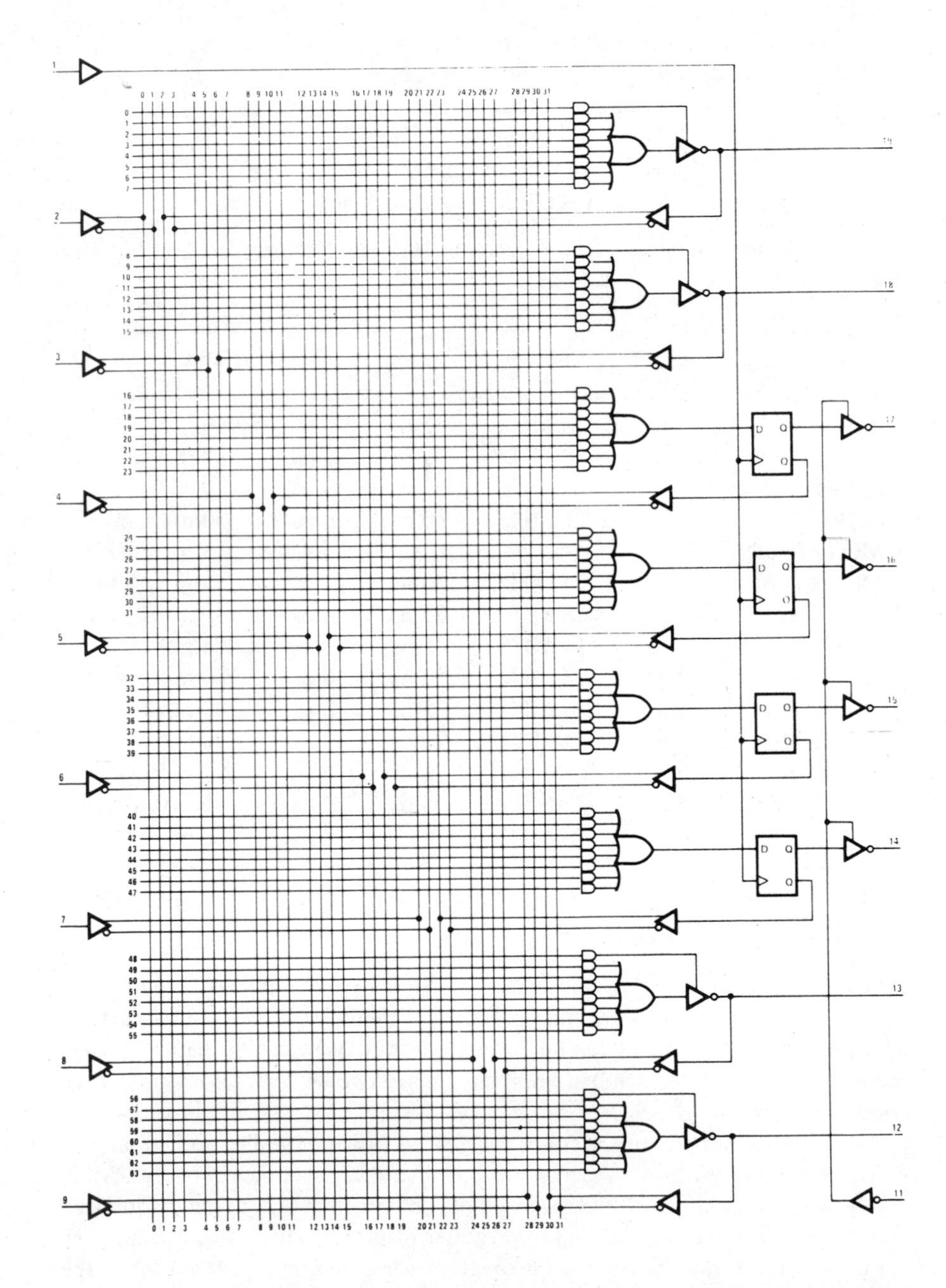

Fig. 5.17 PAL16R4 circuit diagram. *(Reproduced by permission of Monolithic Memories Inc.)*

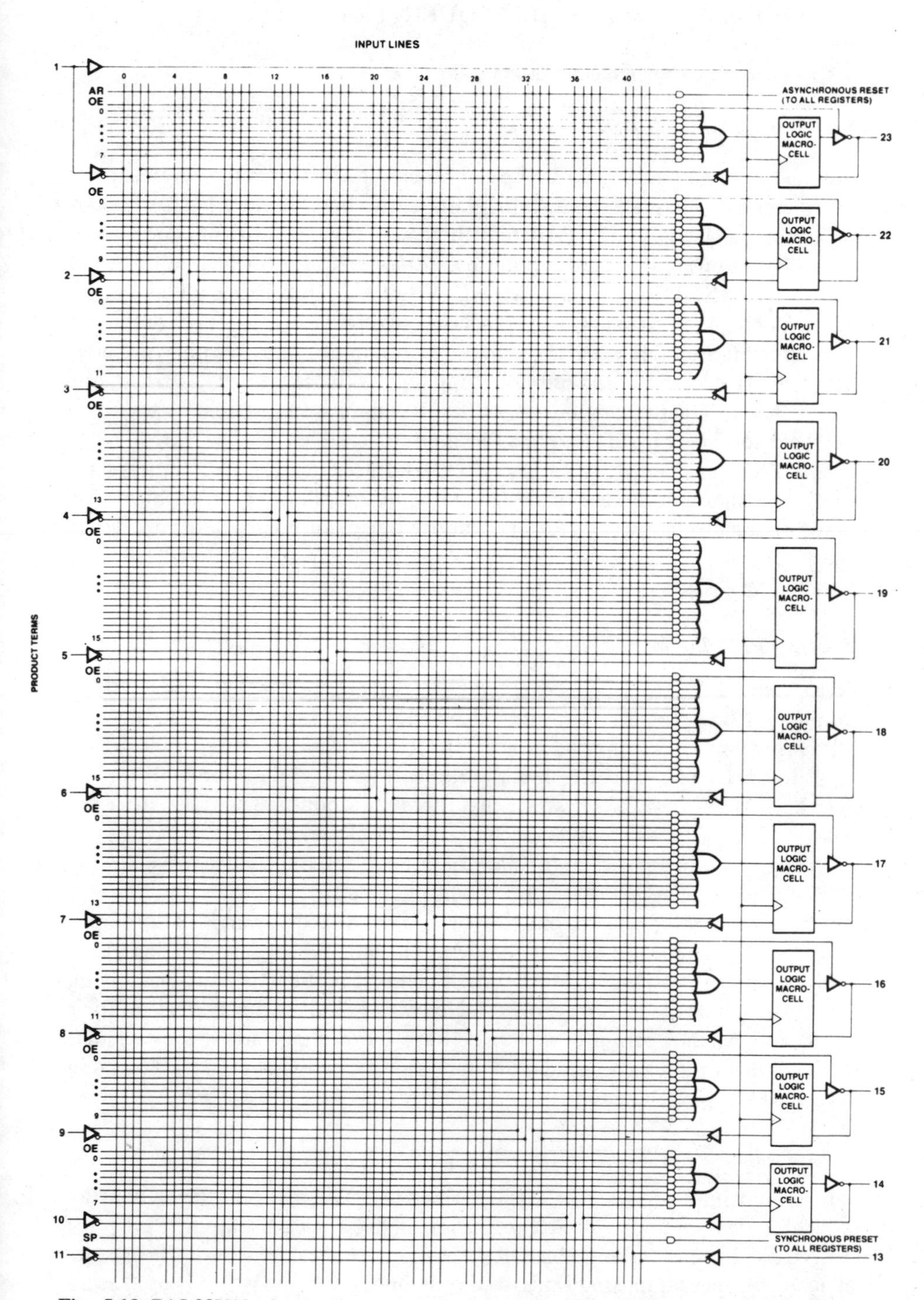

Fig. 5.18 PAL22V10 circuit diagram. *(Reproduced by permission of Advanced Micro Devices)*

5.3 PROGRAMMABLE LOGIC SEQUENCERS

5.3.1 State machine implementation

5.3.1.1 Restrictions of PALs

We have seen how registered PALs can be used to implement state machines. The method was to draw the state diagram, derive the equations for the D-type register and then select the PAL that these equations will fit. This process is necessary because the output register uses D-type flip-flops which must be set whenever a HIGH output is specified. A LOW output is generated by the absence of an input signal. Most PALs have only eight AND gates per output so logic minimisation may be necessary if a complex state machine is being designed. The usual method of minimising logic is to use a Karnaugh map, but these must include the state bits as well as the inputs, so most complex systems will have too many inputs to make this method feasible.

The PAL structure does not therefore lend itself readily to complex state machines, although the use of design software can alleviate the problem if it performs logic minimisation. From the above discussion it may be surmised that the problems are due to the use of D-type flip-flops and the fixed OR-gate structure of PALs.

5.3.1.2 PLS flip-flops

Registered PLAs are called *Programmable Logic Sequencers* or PLSs, which emphasises their potential as state machine devices. There are two families of PLS, which we will examine in more detail later; one family is based on R–S flip-flops while the other one uses J–K flip-flops. If we recall the truth table for the R–S device (Section 3.3.2.4), we see that three operations are possible; these are load HIGH, load LOW and hold. These operations make it possible to miss out the design stage of converting the state diagram to equations since the next state in any sequence can be loaded directly into the register.

The one operation which cannot be carried out directly by an R–S flip-flop is to toggle. J–K flip-flops, on the other hand, include the toggle operation along with the three R–S modes. Most state machines, particularly those involving counting, can be made more economically with J–K flip-flops because of their larger number of operating modes. Perhaps more important is the fact that design is made more easy because the state transitions do not need transforming into AND–OR-type equations.

5.3.1.3 Benefits of the PLS structure

The second disadvantage which we described with PALs was the fixed OR array which limits the number of AND gates available for each output. PLSs, like PLAs, contain a programmable OR gate array. The real benefit of this in state machines is that jump conditions need only one AND gate which is used as an input to the respective OR gate for each flip-flop. This becomes clear when the actual structures are described, with an example in the next section.

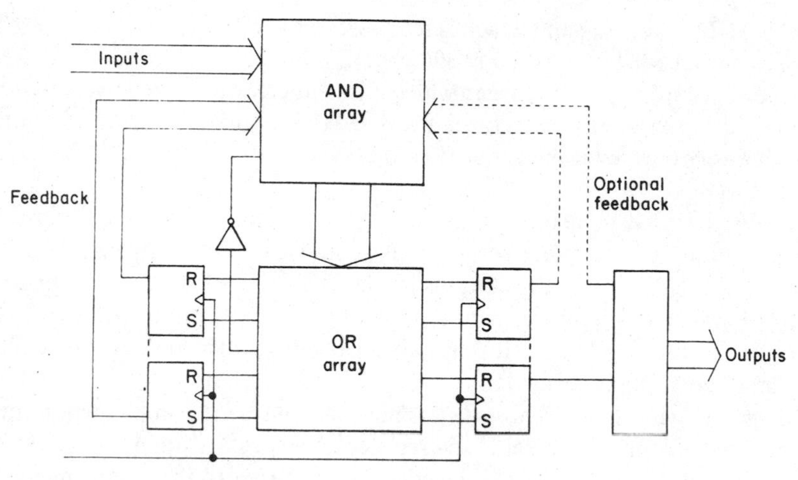

Fig. 5.19 PLS with R–S flip-flops.

5.3.2 PLSs with R–S flip-flops

5.3.2.1 Basic structure

Figure 5.19 shows the structure of the PLS family which uses R–S flip-flops; this family was chronologically the first to be introduced. The AND gate array and OR gate array have the same structure as a PLA, that is the true or complement of any of the inputs can be gated together and the resulting logic function ORed with any of the other AND functions to form an output function. In the case of the PLS these output functions are used to drive either the set or reset of the R–S flip-flops forming the state register.

The state register is divided into two parts: a buried register section which is fully fed back as inputs to the AND gate array, and an output register, only part of which is fed back. The reasoning behind this architecture is that the PLSs in this family have from twelve to sixteen inputs so the state feedback can be extended without compromising the number of inputs too greatly. This architecture is compatible with the Mealy type of state machine in which the output is not tied directly to the internal state of the machine but may, for example, depend on the input as well as the present state. An example would be a keyboard encoder in which the fedback signals are used to perform the scanning and debounce functions; then, having detected a depressed key, it would set the output to the code corresponding to the key.

If all the outputs are fed back to the AND array, then the PLS behaves as a Moore machine, in which the outputs depend only on the present state of the machine. Most counters and simple controllers are examples of Moore machines.

5.3.2.2 Complement term

One part of the PLS which has not appeared in the other logic devices described so far is the complement term. This is a single NOR gate whose inputs are any of the AND functions, and which is fed back to the AND array. As it is a NOR gate its output will be LOW if any of the AND gates driving it is HIGH; in which case it will cause any AND gate to which it is connected to be inactive. However, if all of its inputs are LOW its output will be HIGH and it will allow any gate to which it is connected to be active provided all the other inputs to that gate are true as well.

It may be used to define default conditions within the state diagram. If a number of jumps are defined out of a given state a default jump can be defined if none of the other input conditions is true. This is done by connecting all the defined jumps to the complement term, and gating the output with the present state in a further AND gate. It is usually more efficient to do this than define all the other possibilities individually.

By gating with the present state it is possible to use the complement term for more than one present state. If the register is in state 'A' then any AND gate which includes state 'B' must be inactive so the output of the complement term depends only on whether any of the defined jumps out of state 'A' is true. Even if its output is HIGH the AND gate including the complement term and state 'B' will be LOW and will not affect the outputs from the AND array.

More general use can be made of the complement term as a general way of reducing the number of AND gates needed for all the jumps from a given state. When we were looking at the Karnaugh map in relation to combinational logic we saw that programmable outputs allowed us to group either '1's or '0's and choose the lower number of groups. If we draw the Karnaugh map for just the input conditions for each state jump we must group the '1's if we are to drive the flip-flops directly. However, if we use the complement term, we can group the '0's to give us the same function. This can be used as the solution if fewer AND gates are required, remembering that an extra AND gate is required to gate the complement with the present state.

The only drawback with this technique is that the extra feedback path introduces an extra delay into the logic path. It is not applicable, therefore, to designs where speed is at a premium. As an indication of the effect of this extra path, the set-up time is increased by about 30 ns, while operating frequency can be reduced from 20 to 12.5 MHz.

5.3.2.3 Design example

The design method for PLSs is best explained by reference to a worked example; we can use the enhanced combination lock again as an illustration. As we can bypass the logic equation stage we can go back to the state diagram and translate this directly to a state table. Referring to Figure 3.38, where we originally introduced the state diagram, we can define all the state jumps directly as:

Inputs				*Present State*				*Next State*				*Op*
I3	I2	I1	I0	Q3	Q2	Q1	Q0	Q3	Q2	Q1	Q0	UN
H	L	L	L	L	-	L	L	L	-	L	H	L
L	L	L	L	L	-	L	H	L	-	H	L	L
H	L	L	H	L	-	H	L	L	-	H	H	H

These three lines of state table define all the 'legal' jumps in the state diagram. If the three AND gates are also connected to the complement term we can use this to define the jumps for incorrect entries. As long as one of the input/present state combinations is true the complement term will have a LOW output. If an incorrect entry is made then none of the inputs will be HIGH so the complement output will be HIGH. If this is gated with the 'first mistake' bit, Q2, the incorrect entry will set either Q2, if this is the first mistake, or Q3 if it is the second.

The convention for incorporating the complement term into a state table needs some explanation. In principle it should appear in both the output and input sides of the state table as it can take on either role. In practice it is given a separate column on the input side with an 'A' used to define it as an output (generate) and a '.' as an input (propagate). If it is not being used as either, it should be programmed '–' (don't care) so that it does not interfere with normal operation. The full state table then appears as:

Inputs					*Present State*				*Next State*				*O/P*
C	I3	I2	I1	I0	Q3	Q2	Q1	Q0	Q3	Q2	Q1	Q0	UN
A	H	L	L	L	L	–	L	L	L	–	L	H	L
A	L	L	L	L	L	–	L	H	L	–	H	L	L
A	H	L	L	H	L	–	H	L	L	–	H	H	H
.	–	–	–	–	L	L	–	–	L	H	–	–	L
.	–	–	—	–	L	H	–	–	H	H	–	–	L

Note that in this design we have specified UN as the 'unlock' function and have generated this synchronously as a registered output which plays no part in the state jumps. PLSs in this family can be given an asynchronous preset which would set the register to the complement of what we require from the RESET we specified previously. The registers also power-up in a state with all HIGHs so a practical solution would need the state bits exactly the inverse of the way we have specified them. That this would not affect the operation of the circuit at all goes to show the versatility of this type of device. It is also remarkable that a circuit with nine states and twelve transitions can be specified in just five AND terms!

The other important benefit of this structure is that we were able to enter the states directly into the state table, which is a format accepted by most commercial programming equipment. This also allowed us to perform logic minimisation by inspection, by recognising that the 'legal' jumps are independent of the 'first mistake' bit, Q2, which can therefore be set as don't care. In a more complex system we might have to use one of the formal techniques already described if the logic will not fit into the PLS.

5.3.3 PLSs with J-K flip-flops

5.3.3.1 The composite flip-flop

Although this section is headed 'PLSs with J–K flip-flops' the other PLS family uses a flip-flop surrounded by many frills which enable it to perform many

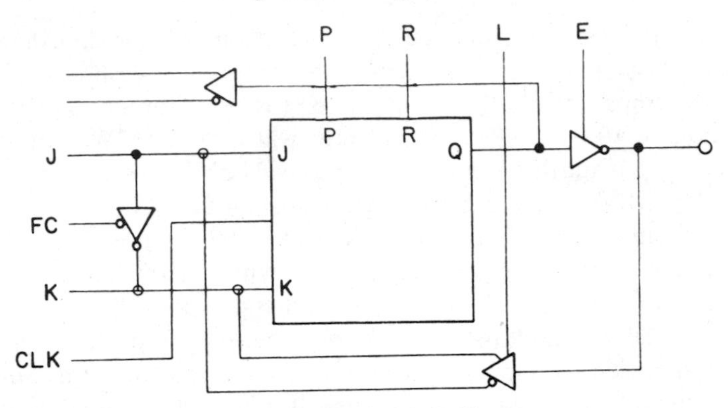

Fig. 5.20 'Composite' J–K flip-flop.

functions, in addition to those normally associated with a basic flip-flop. The circuit diagram of the composite flip-flop is shown in Figure 5.20.

The first addition is a programmable inverter driving the 'K' input from the 'J' input. Provided that the 'K' input is otherwise disabled, this feature makes the flip-flop behave in the same way as a D-type, because the conditions J=K=LOW and J=K=HIGH cannot occur. If 'J' is HIGH then 'K' is LOW so a HIGH is loaded, while a LOW on 'J' sends 'K' HIGH to load a LOW. This may seem a retrograde step as the D-type has less functions than a J–K, but there are applications where just loading is required, as in registers and pipeline circuits, when the D-type is more efficient. The inverters are controlled by a separate fuse array and AND gate so that the type of flip-flop can be preset, or changed during circuit operation.

Also connected to the 'J' and 'K' inputs are the outputs from a true/complement buffer driven from the device outputs. This is the same as the register preload described in connection with some of the more complex PALs. Preload should only be used when the outputs are disabled by the tri-state control to avoid contention on the output pins. As with PALs, this facility may be used to assist in testing or to ensure proper start-up of a state machine. Another possibility is that the output pins can be connected to a data bus, the register loaded with data which is modified according to the input conditions and loaded back on to the bus. The tri-state control operates after the feedback into the AND array so the PLS can be hidden from the bus until its data is ready to be read.

As well as these two novel features, the composite flip-flop has asynchronous preset and reset making it a very versatile circuit element.

5.3.3.2 Overall structure

Having described the engine of this family of PLSs, we are now in a position to review its overall structure as illustrated in Figure 5.21. It is, in effect, an extension of the PLS153 architecture. The unregistered section of the PLS is a reduced version of the PLS153 with a set of inputs and bidirectional pins, and an

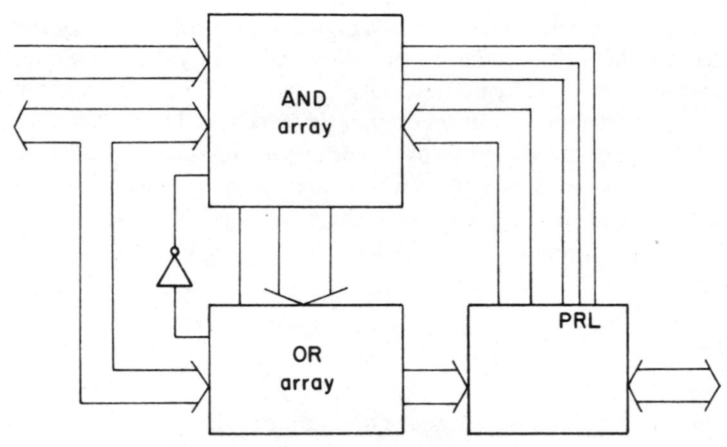

Fig. 5.21 PLS with J–K flip-flops.

array of AND gates driving the programmable OR-array. As with registered PALs, this makes it possible to mix registered and unregistered functions in the same device, or dedicate all the unregistered pins as inputs.

Unlike the R–S-type PLSs, there is no buried register in the J–K family. All the flip-flops are connected to output pins and are also fed back to the AND-gate array, so these PLSs can be used to build Moore-type state machines. The gates which drive the preset, reset and parallel load functions of the composite flip-flops are located alongside the tri-state control gates for the unregistered bidirectional pins. All these gates form a control section of the PLS; it is fruitful for the designer to keep this distinction clear in his mind since it makes the design process more rigorous.

The only control function which does not reside in this area is the flip-flop-type control. This is the term which defines the behaviour of the flip-flop; whether it acts as a J–K or as a D-type. Each of the J-to-K inverters has a fused connection to an AND-gate output called the flip-flop control term. It is logically connected so that the inverter is enabled by a LOW on its control input. Because the AND gate has a LOW output in its unblown state it follows that the flip-flops will behave as D-types if all the fuses are left intact. To achieve J–K operation it is necessary to ensure a HIGH on the control input. This will be the case if the control fuse is blown, or if the flip-flop control term is at a HIGH level itself.

The other features which this family shares with the R–S family are the complement term and a tri-state enable for the registered outputs. We have already dealt with the use of the complement term but two more points are worth making regarding its use in this family. Firstly, it may be used with the unregistered outputs as an internal feedback to make, for example, an exclusive-OR function without using up outputs. Also, its output may be connected to the control section. A possible use might be to force a reset if an illegal state is detected.

The tri-state enable has to be used in conjunction with the parallel load function. If this function is not being used and tri-stating is not required it is possible to permanently enable the outputs by blowing the appropriate fuse, as detailed in the device data sheets. One possibility which should not be overlooked is that the registered outputs could be used as registered inputs instead by leaving the outputs permanently tri-stated. The device can be used as a PLA with registered inputs by treating the bidirectional pins as the only outputs. This technique is used when interfacing asynchronous signals to a synchronous system; using a PLS has the advantage that logic functions can be included in this *staticiser*, as it is called.

5.3.3.3 Design example

Once again we can best illustrate the design procedure by means of an example. Let us again build a four-bit counter but this time extend the function so that we have the option of a binary count or Gray code counting. We have already encountered Gray code in making Karnaugh maps; in Gray code only one output at a time toggles so possible problems due to dynamic hazards are avoided when decoding the count. Firstly, though, we will see how to build a binary counter from the J–K family, and then design the additional Gray code counter.

In the discussion on registered PALs in Section 5.2.1.5 we saw how to build a counter with D-type flip-flops, and how the number of terms for each bit escalated as we progressed up the counter bits. This was due primarily to the need to define the hold conditions. J–K flip-flops have the hold condition built in so, without analysing the situation formally, it seems likely that J–Ks are going to provide a simpler solution than D-types. In fact, all we have to define is that each bit toggles when all the lower order bits are HIGH. A J–K toggles when J=K=HIGH so we need to program the following equations:

Q0: J = K = BIN
Q1: J = K = BIN * Q0
Q2: J = K = BIN * Q0 * Q1
Q3: J = K = BIN * Q0 * Q1 * Q2

BIN is a signal which we use to indicate that we are counting in binary. In order to convert them into a state table we must see how the flip-flop inputs are specified. Each flip-flop has two inputs, ‘J’ and ‘K’ and we can write up a truth table relating fuse conditions to the functions and table entry convention:

‘J’	*‘K’*	*Code*	*Function*
intact	intact	0	toggle
intact	blown	H	load HIGH
blown	intact	L	load LOW
blown	blown	–	hold

Using these conventional symbols we can generate the following state table:

		Inputs				*Outputs*		
Bin	*Q3*	*Q2*	*Q1*	*Q0*	*Q3*	*Q2*	*Q1*	*Q0*
H	–	–	–	–	–	–	–	0
H	–	–	–	H	–	–	0	–
H	–	–	H	H	–	0	–	–
H	–	H	H	H	0	–	–	–

We may now design the Gray code half of the counter. The sequence of counting is shown in Figure 5.22; one way to put this into a PLS would be to enter each transition as a separate AND term. This would use sixteen AND gates, so let us investigate the possibility of reducing this number by examining the transitions of each bit in turn. Since we have the option of using D-types or J–K flip-flops we can draw the Karnaugh maps for both loading HIGHs and toggling. The results are shown in Figure 5.23. The number of AND gates required for each map is indicated by the map.

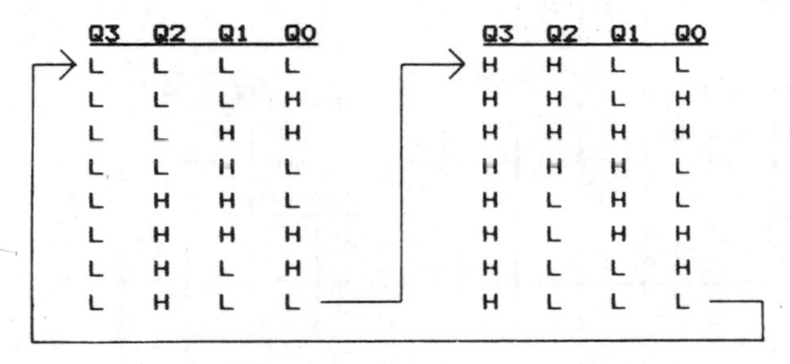

Fig. 5.22 Gray code count sequence.

Not surprisingly, using J–K flip-flops exclusively would take sixteen gates because the property of Gray code is that the bits toggle one at a time when counting. The appearance of the maps suggests that the exclusive-OR function would be useful; for example, we can write Q0 as:

$\overline{Q0}$:= Q3 :+: Q2 :+: Q1

However, this would take more gates to build than merely entering the four gates directly from the map, even if we use the complement term. On this basis we could build the Gray counter from thirteen gates using D-type flip-flops, by reference to the Karnaugh maps. More careful analysis shows that we can save two more gates by using J–Ks for Q3 and Q2, and D-types for Q1 and Q0. The equations for this section are, therefore:

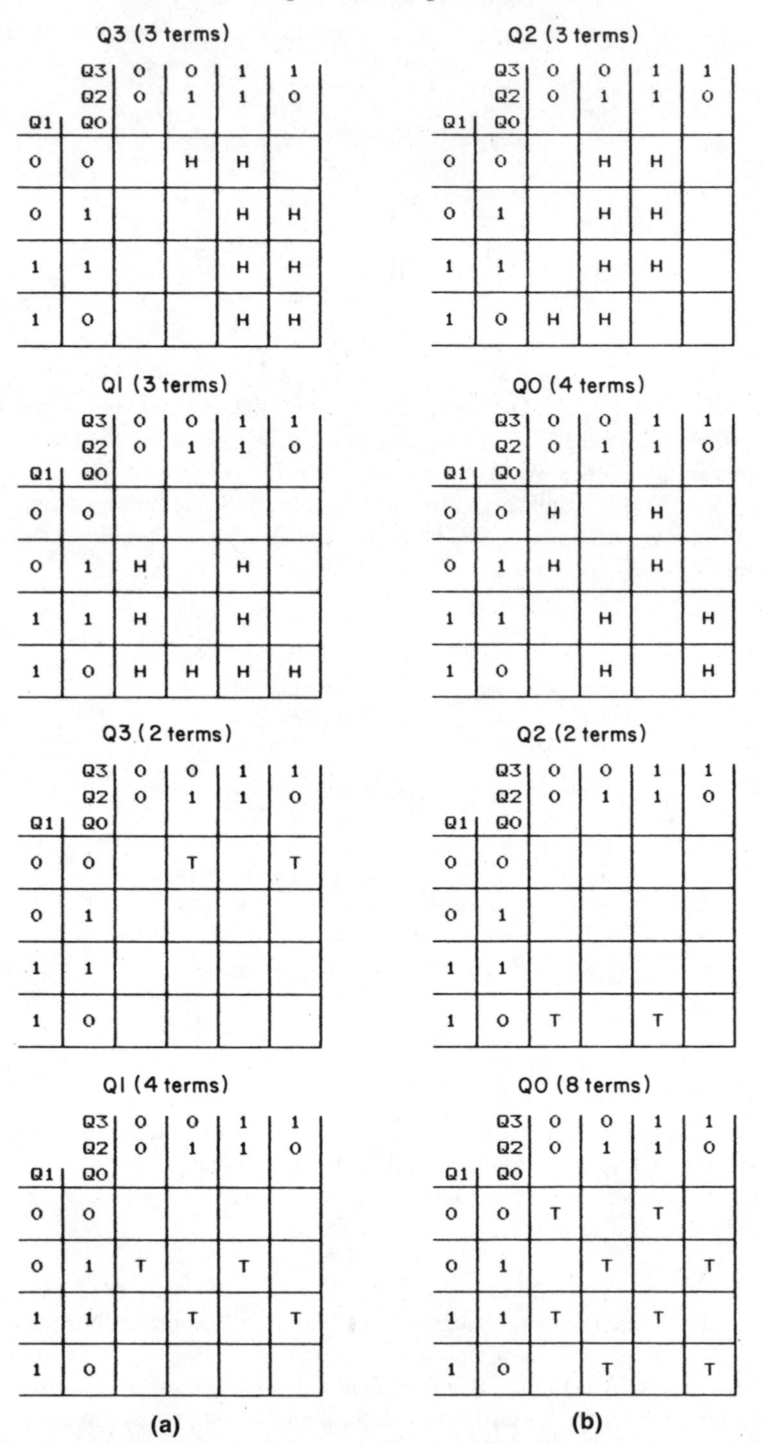

Fig. 5.23 Karnaugh maps – Gray code counter. (a) Specifying 'H'; (b) specifying 'toggle'.

$$
\begin{aligned}
Q3\text{: } J = K &= \overline{BIN} * \overline{Q3} * Q2 * \overline{Q1} * \overline{Q0} \\
&+ \overline{BIN} * Q3 * \overline{Q2} * \overline{Q1} * \overline{Q0} \\
Q2\text{: } J = K &= \overline{BIN} * \overline{Q3} * \overline{Q2} * Q1 * \overline{Q0} \\
&+ \overline{BIN} * Q3 * Q2 * Q1 * \overline{Q0} \\
Q1 &:= \overline{BIN} * \overline{Q3} * \overline{Q2} * Q0 \\
&+ \overline{BIN} * Q3 * Q2 * Q0 \\
&+ \overline{BIN} * Q1 * \overline{Q0} \\
Q0 &:= \overline{BIN} * \overline{Q3} * \overline{Q2} * \overline{Q1} \\
&+ \overline{BIN} * \overline{Q3} * Q2 * Q1 \\
&+ \overline{BIN} * Q3 * Q2 * \overline{Q1} \\
&+ \overline{BIN} * Q3 * \overline{Q2} * Q1
\end{aligned}
$$

To complete the design we have to define the control terms needed to configure the flip-flops. Q3 and Q2 are J–Ks in both cases so we can blow the fuses to detach them from the flip-flop control line; Q1 and Q0 are J–Ks when BIN is HIGH and D-types when BIN is LOW so the flip-flop control equation is:

FC = BIN

All this information can be entered into a programming table as under:

	I/P	*Present State*					*Next State*		
	Bin	*Q3*	*Q2*	*Q1*	*Q0*	*Q3*	*Q2*	*Q1*	*Q0*
					F/F	.		A	A
	H	–	–	–	–	–	–	–	0
	H	–	–	–	H	–	–	0	–
	H	–	–	H	H	–	0	–	–
	H	–	H	H	H	0	–	–	–
	L	L	H	L	L	0	–	–	–
	L	H	L	L	L	0	–	–	–
	L	L	L	H	L	–	0	–	–
	L	H	H	H	L	–	0	–	–
	L	L	L	–	H	–	–	H	–
	L	H	H	–	H	–	–	H	–
	L	–	–	H	L	–	–	H	–
	L	L	L	L	–	–	–	–	H
	L	L	H	H	–	–	–	–	H
	L	H	H	L	–	–	–	–	H
	L	H	L	H	–	–	–	–	H
FC	H	–	–	–	–				

There are a number of points which come out of the above table. The 'F/F' term defines which fuses connecting the flip-flops to the 'FC' line are to be left intact; as with the OR-array an 'A' denotes an intact fuse and '.' a blown fuse. The only valid next state entries for a D-type are 'H' and '–', otherwise there would be contention between the 'K' OR gate and the J-to-K inverter. Because we are not using the complement term or a preset or reset, these have been omitted for the sake of clarity. In a full table entry the complement term would have '–' in every line to remove it from the array; '0' (both fuses intact) is an illegal entry for the complement term as it could lead to oscillation. In order to make the other functions inactive we would leave all fuses intact, that is, '0' in all entries in the programming table.

5.3.4 PLS availability

5.3.4.1 Technology and performance

At present, PLSs are available only in bipolar technology although there are plans to introduce some architectures in CMOS. As with PLAs, the programmable OR-array means that there is some reduction in speed performance when compared with the simple registered PALs. This is due, in part, to the capacitative loading effect on the OR gates; in PALs, no more than eight AND gates are connected to any one OR gate. Some of the PLS data sheets indicate how the performance can be improved by restricting the OR connectivity. The limits quoted below assume that no more than twelve AND gates are connected to any OR gate; any increased connectivity can be adjusted for by reference to the appropriate data sheet.

The other factor affecting performance is the design of the flip-flops. The D-type flip-flop, as used in PALs, is basically a faster structure than the J–K flip-flop, so it will yield an intrinsically faster device. We have discussed above the advantages that using J–Ks brings in terms of easier design and more complex functions. In deciding whether to use a PAL or a PLS it is necessary to weigh the benefits of higher potential speed against a more powerful architecture. The performance of the two PLS families may be summarised as below:

Family	*Delay*	*Clock*	*Set-up*	*Hold*	*Supply*
R–S-Type	20 ns	20 MHz	30 ns	5 ns	180 mA
J–K-Type	30 ns	15 MHz	40 ns	0 ns	190 mA

5.3.4.2 Architecture details

The following table lists the various PLS devices available and gives details of their internal structures.

Part Number	*F/F Type*	*Inputs*	*B I/O*	*Register Details* *Buried*	*F/B*	*Output*	*AND Gates*
PLS105	R–S	16	0	6	0	8	48
PLS155	J–K	4	8	0	4	0	32
PLS157	J–K	4	6	0	6	0	32
PLS159	J–K	4	4	0	8	0	32
PLS167	R–S	14	0	8	0	6	48
PLS168	R–S	12	0	6	4	4	48
PLS169	R–S	16	0	6	0	4	48

The full circuit diagrams of the PLS105 and the PLS159 are shown in Figures 5.24 and 5.25 respectively.

5.4 PLDs WITH OTHER ARCHITECTURES

5.4.1 Asynchronous PALs

5.4.1.1 Drawbacks of traditional structure

Many circuits designed in standard logic rely on flip-flops clocked from different sources for their basic operation. To design a PLD to replace these circuits means either building the flip-flops from discrete AND–OR gates in a combinational PLD, or redesigning the function to build it from a synchronously clocked PLD. Other designs might need some flip-flops to be reset and others left unchanged by a reset signal. This again is a problem, as most PLDs have a universal set and reset which applies to all flip-flops, or none. For example, Figure 5.26 shows a simple circuit for generating a waveform which is HIGH for 5/2 clock pulses and LOW for 3/2 pulses. To make this function from a standard registered PLD would require a clock of double the frequency and an extra flip-flop.

5.4.1.2 The asynchronous PAL cell

In order to produce a range of PALs with a more flexible structure, including asynchronous clocking and individual flip-flop control, an asynchronous cell has been developed. The term *State-Machine Atomic Cell* or SMAC (a trademark of MMI) has been coined for this feature, which is shown in Figure 5.27. It contains eight AND-terms, but only four of these are combined to form a standard logic group. The other four provide a clock, set, reset and output enable which are dedicated to the flip-flop and output from that cell. By taking set and reset both HIGH, a condition which is normally illegal, the register is bypassed to convert the cell into a combinational output. All the outputs are fed back to the AND-array, so they are in effect bidirectional pins.

A device is constructed by putting together as many cells as will fit into the package, the limitation being the number of inputs and outputs. Note that each cell uses one input and one output. In addition to the individual cell features there is a global output enable and parallel load. Thus a twenty-pin package can accommodate eight cells, and a twenty-four-pin package ten cells. The provision of dedicated and global output enables means that each bidirectional pin can be set as an input or an output, or controlled on an individual basis, while every output can also be turned off simultaneously. The latter feature is necessary when using the preload in testing to set the flip-flops to a known state, apart from any possible use in normal device operation.

A slightly different cell design is shown in Figure 5.28; this is used in the EP series of PALs. There are ten AND terms associated with each I/O cell, eight of these are used for creating the logic function, one is a clear which takes the flip-flop LOW and the tenth can be used as either a clock or output enable. A

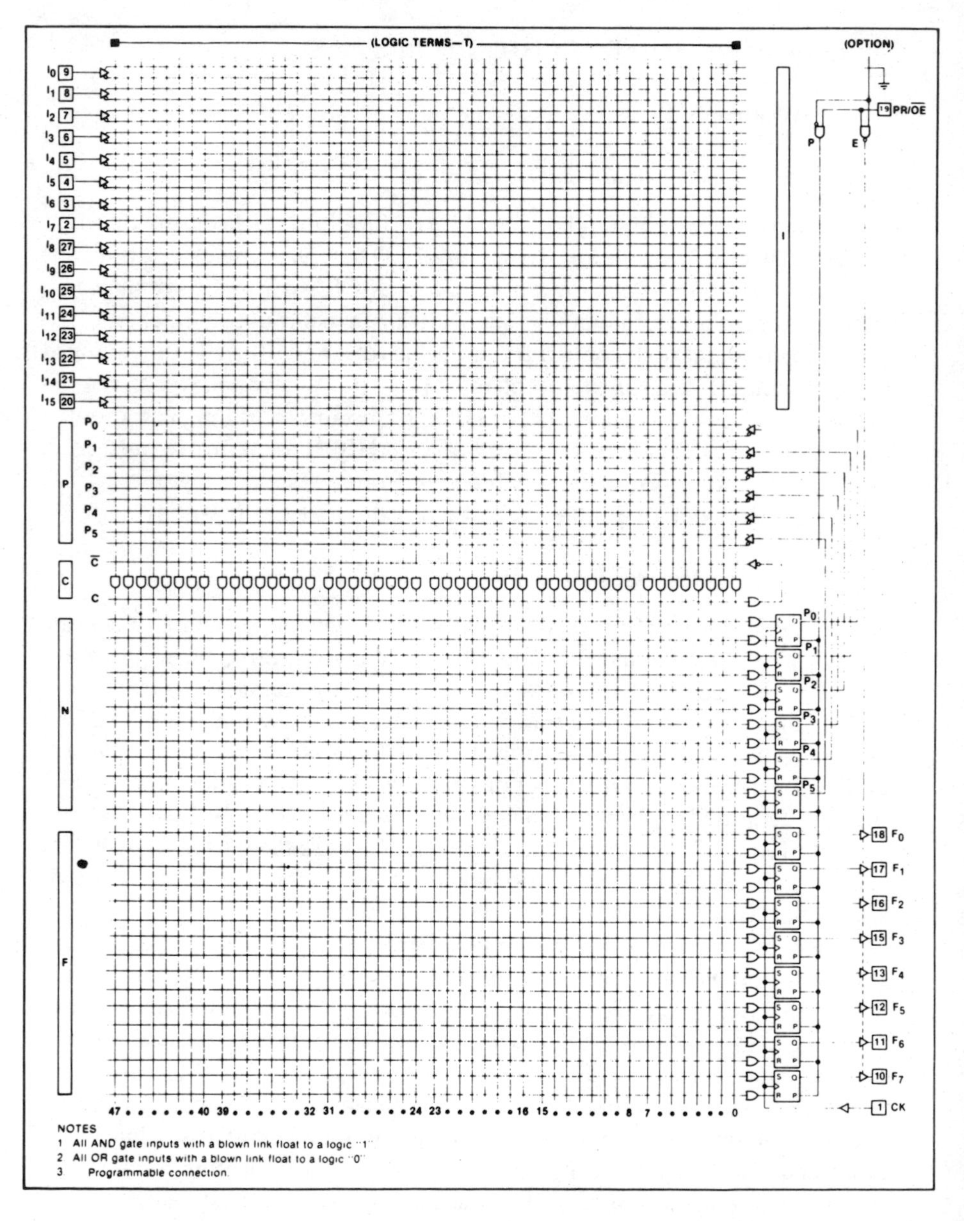

Fig. 5.24 PLS105 circuit diagram. *(Reproduced by permission of Mullard Ltd)*

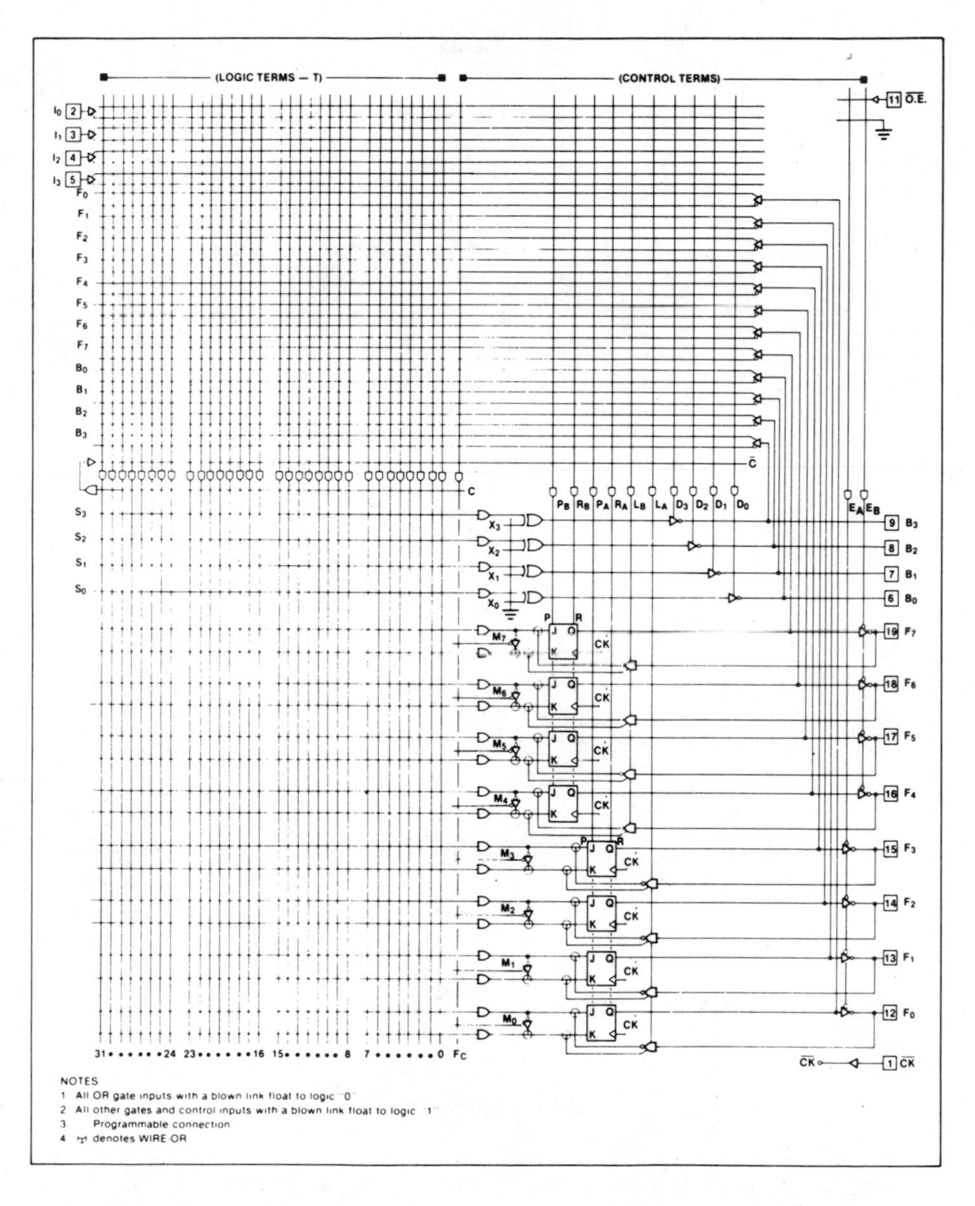

Fig. 5.25 PLS159 circuit diagram. *(Reproduced by permission of Mullard Ltd)*

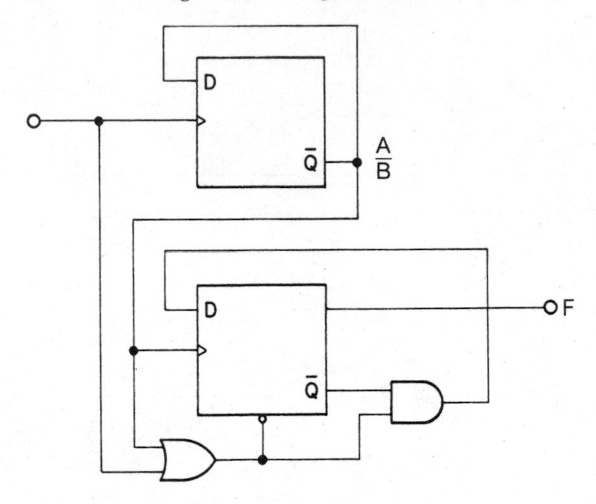

Fig. 5.26 Waveform generator with 5/3 mark/space ratio.

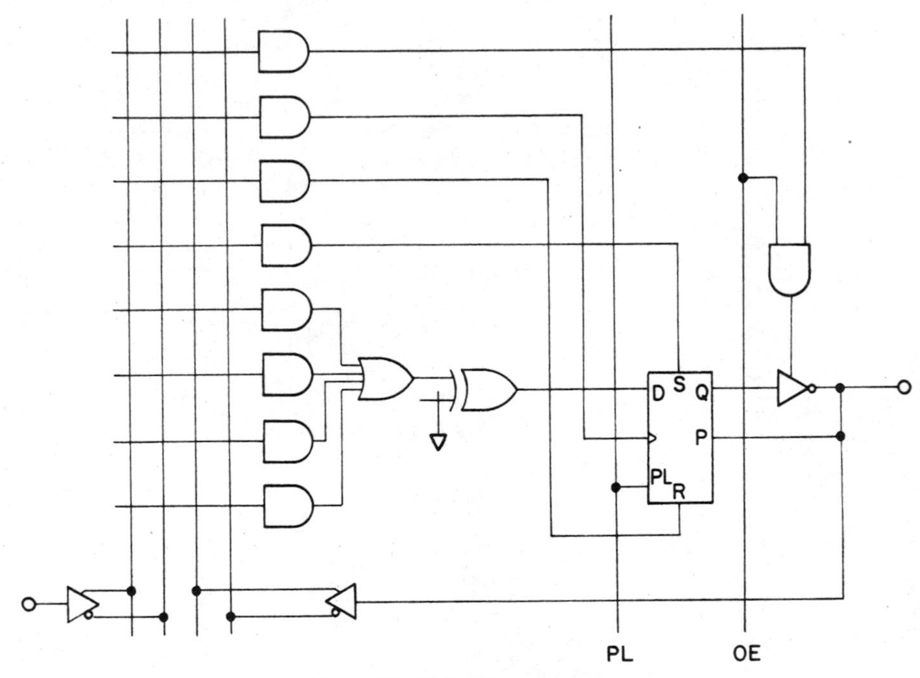

Fig. 5.27 'SMAC' structure.

programmable multiplexer allows the clock to be taken from either the AND term or a global clock line giving the option of either individual clocking or a common clock. Similarly, the output enable can be either permanently ON or driven from the AND term. This arrangement is a little more restricting than the SMAC as there is no preload facility and no global output enable, although these could be provided via the logic array if a common clock is being used.

The chief advantage of the EP series output macrocell is its versatile choice of output type. The architecture bits may be programmed to provide almost any flip-flop configuration, or an unregistered output. Designs can be made with D-type, J–K, R–S or toggle flip-flops, and feedback selected from either the flip-flop or the pin for the D-type and toggle options. In J–K and R–S modes, the AND terms are shared between the two flip-flop inputs.

5.4.1.3 Design example

In the SMAC architecture, half of the AND terms are dedicated to specific functions so a particular convention is used to define the function in the logic equations. We can see how the convention is applied if we write the equations for the 5–3 mark–space waveform generator from Figure 5.26.

$$B := \overline{B}$$
$$B.CLK = A$$

$$F := A * \overline{F} + \overline{B} * \overline{F}$$
$$F.CLK = \overline{B}$$
$$F.SET = \overline{A} * B$$

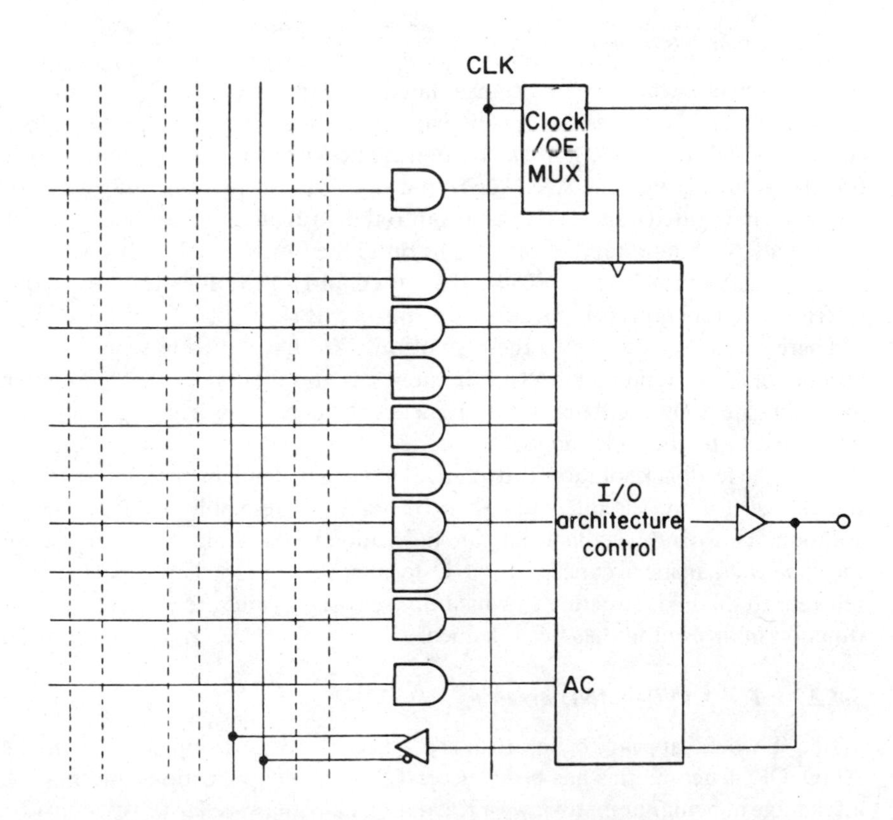

Fig. 5.28 EP-series asynchronous PAL cell.

5.4.1.4 Asynchronous PAL availability

The following table is a summary of the structure and performance of the currently available asynchronous PALs:

Number	*Inputs*	*I/O*	*Cell*	*f.max*	*Delay*	*Supply*
16RA8	8	8	SMAC	20 MHz	30 ns	170 mA
20RA10	10	10	SMAC	20 MHz	30 ns	200 mA
EP600	4	16	EP ser	30 MHz	25 ns	4 mA @ 1 MHz
EP900	12	24	EP ser	25 MHz	30 ns	6 mA @ 1 MHz
EP1800	12	48	EP ser	16 MHz	50 ns	20 mA @ 1 MHz

As may be inferred from the power supply figures, the SMAC devices are currently fabricated in bipolar technology, while the EP series is CMOS. The full circuit diagram of the 20RA10 is shown in Figure 5.29.

5.4.2 Programmable macro logic

5.4.2.1 Architecture principles

All the devices which we have examined so far have been based on an architecture of AND gates driving OR gates. The principle behind this is that any combinational logic function can be represented on a Karnaugh map which translates directly into the AND–OR structure. Most of the functions we have discussed have fitted quite economically into this structure, that is, they have not needed an exorbitant number of gates to build the function. Those which need too many gates for a PAL or PLA can be fitted into a PLE or can be built from intermediate functions using feedback terms in a PAL or PLA.

There are disadvantages to these solutions. We have already seen that the number of inputs is limited in PLEs, because an extra input doubles the number of AND gates. The feedback solution incurs the penalty of using up I/O pins when access to the function is not needed elsewhere in the system. Clearly, though, the feedback solution is attractive, if it can be implemented internally to the device, for the circuit could then be made to resemble a discrete logic solution. This would enable total gate utilization to be minimised, rather than the somewhat artificial concept of AND-term minimisation. This is particularly relevant to the PAL structure in which unused AND gates are wasted once the function of any output has been defined.

5.4.2.2 The NAND–NAND array

While it would appear to be attractive to break away completely from the AND–OR structure, this has been successful in many applications and has the advantage of being compatible with Karnaugh map analysis. However, consider the following logic manipulation:

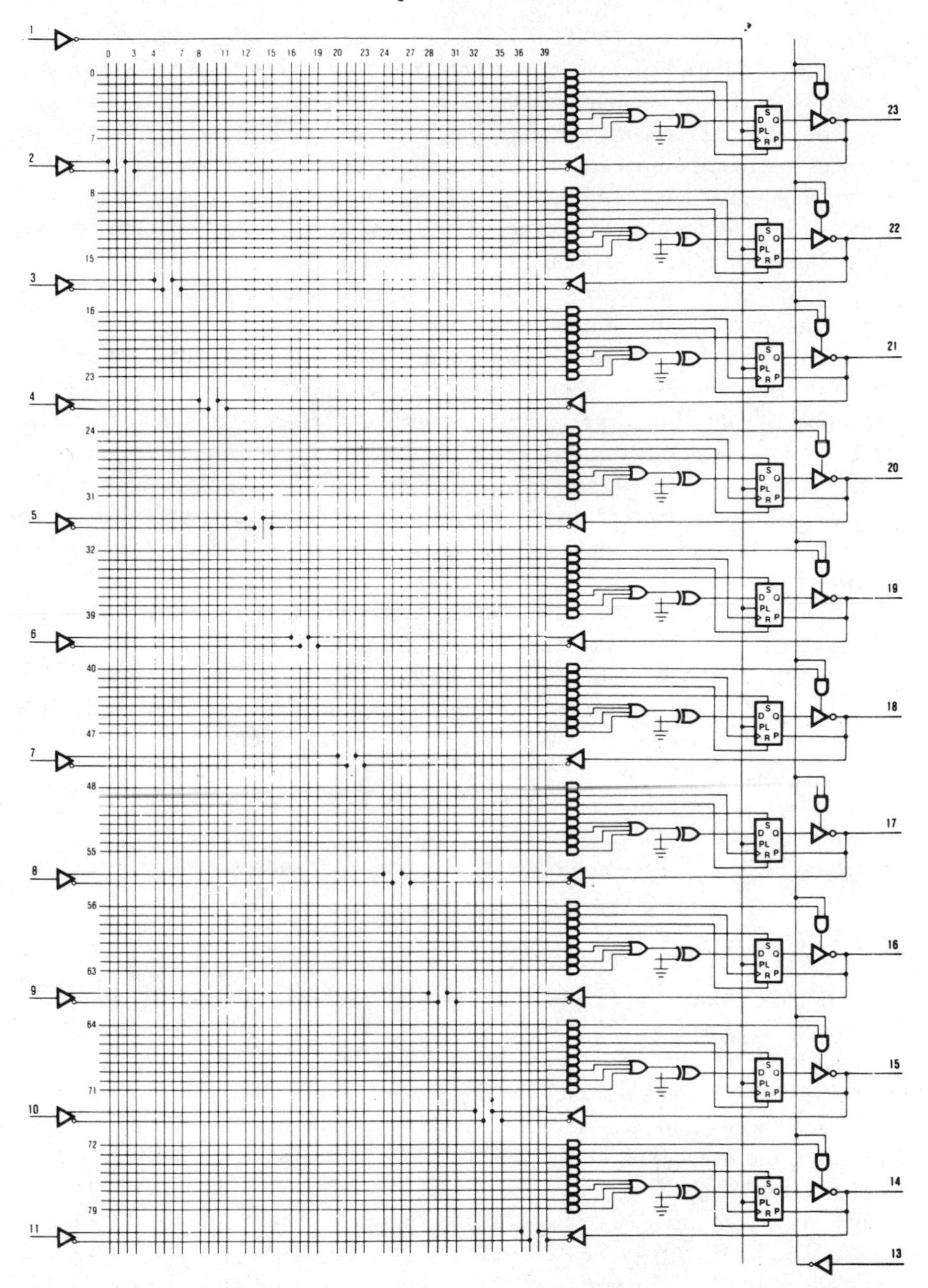

Fig. 5.29 PAL20RA10 circuit diagram. *(Reproduced by permission of Monolithic Memories Inc.)*

$$F = A * B + C * D$$

$$\overline{F} = \overline{((A * B) + (C * D))}$$

$$= (\overline{A * B}) * (\overline{C * D})$$

Thus, AND–OR and NAND–NAND are logically equivalent expressions, but the latter has the advantage that it may be built from only one type of gate. It also means that similar techniques to those used in designing masked gate arrays can be adopted for *Programmable Macro Logic*, or PML. One of the techniques is the use of macros; these are logic functions built from the basic gates which can be used as a higher-level component, in the same way that sub-assemblies are used in mechanical designs. This principle is expanded in the next chapter, but it is extremely suitable for application to this structure.

Figure 5.30 shows the basic NAND array; each 'solid' crossover represents a programmable connection, so each NAND gate takes an input from one of the vertical lines and has a programmable connection from its output to the input of every other gate. The number of fuses increases, therefore, as the square of the number of gates. In order to prevent the number of fuses getting out of hand for the more complex devices, the basic array can be surrounded by dedicated functions such as flip-flops, exclusive-OR gates, arithmetic circuits, memories, etc. These function macros can be completely internal, just as the buried registers found in the more complex PLSs and PALs, and do not use up any I/O capability.

In addition to the internal macros, provision must be made to connect the array to the outside world. This is done with interface macros which can take on any of the functions normally found in PLD interfaces, that is, inputs, I/Os, registered outputs, exclusive-OR outputs, etc. The generalised PML structure would appear as in Figure 5.31 with the NAND array acting as logic glue to the surrounding macros. We have therefore progressed to a point where the glue has ceased to be the primary function of the PLD and has become a secondary feature within the PLD. This is to be expected as device complexity increases to the level where PLDs can include whole systems or subsystems.

5.4.2.3 Proposed device structures

So far, two devices have been defined in the PML family; they are the PLHS501 (Figure 5.32), and the PLHS502 (Figure 5.33). The PLHS501 has no function macros, all the logic being defined in the NAND array. The device has twenty-four dedicated inputs with eight I/Os, giving a possible total of thirty-two inputs. There are four standard outputs arranged in two pairs with a common output enable, and four other outputs are provided; these are exclusive-OR gates with a common enable. If the exclusive-OR function is not required it may be ignored by leaving the input to one half unprogrammed. Finally, the logic content is provided by seventy-two NAND gates.

The connectivity between inputs, logic and output is controlled entirely by the NAND array, where any signal with an input to the array can be connected to any component with an input from the array. This completely free structure means that over 13000 fuses are needed. If we wish to incorporate flip-flops into a design, this can be done by configuring the internal gates as flip-flops. It will take six gates for each D-type or four gates for each R–S flip-flop required; this will use a majority of the gates needed for making the logic connections, so it is

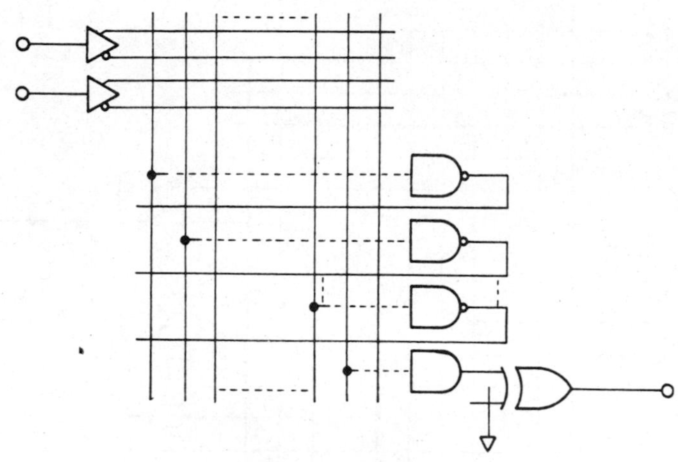

Fig. 5.30 PML NAND array.

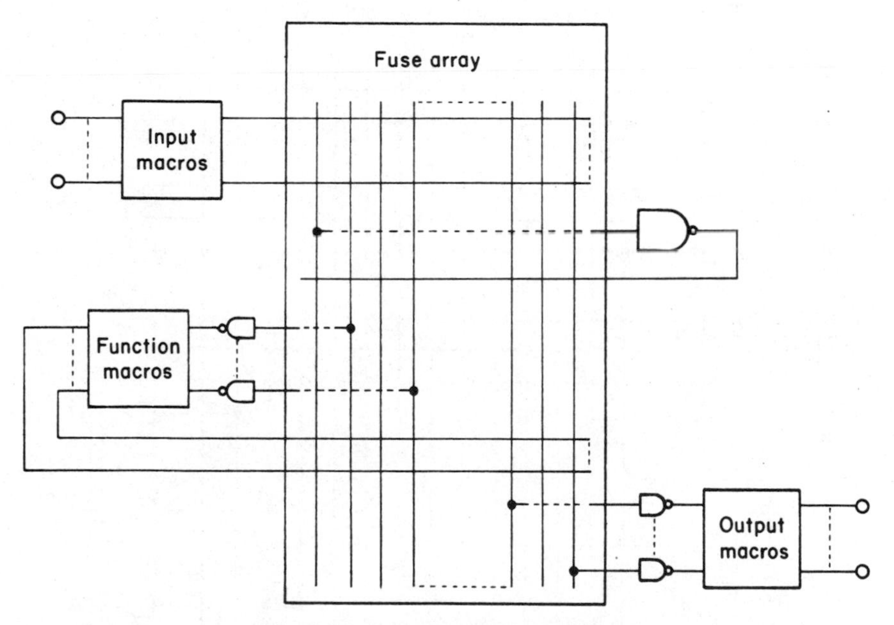

Fig. 5.31 Generalised PML structure.

not to be recommended. To expand the NAND array to leave the same number of gates available for logic would increase the number of fuses to over 27000. Adding flip-flops as functional macros will still increase the number of fuses, but only by about 20 instead of 100 per cent.

This course has been taken in the PLHS502. Some saving has been made by eliminating the exclusive-OR gates and cutting the number of NAND gates to sixty-four, otherwise the structure is the same as the PLHS501. The detailed differences are in connection with the flip-flops themselves. There are two banks

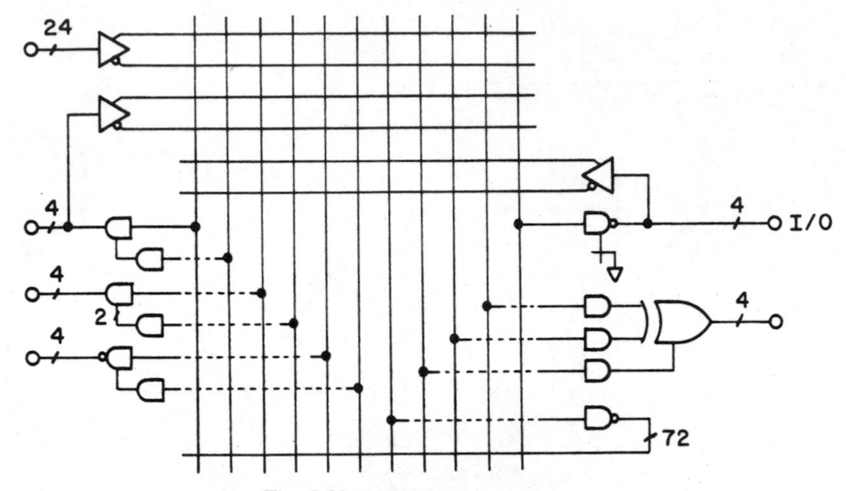

Fig. 5.32 PLHS501 block diagram.

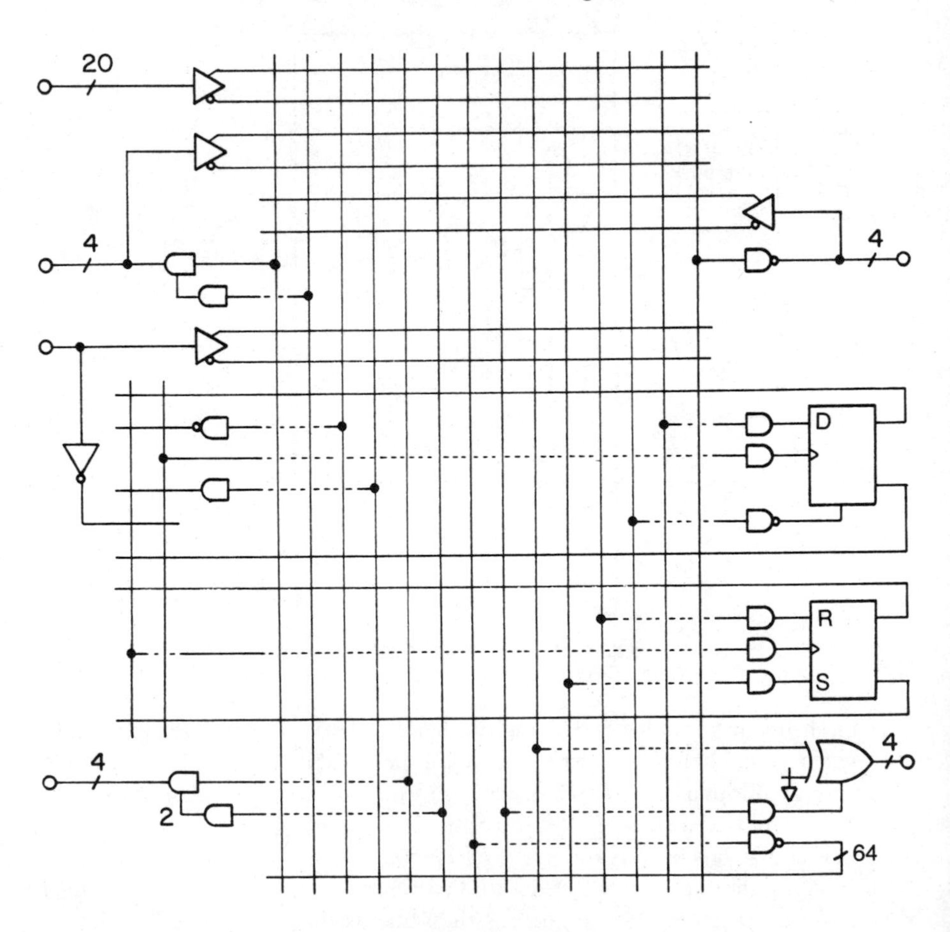

Fig. 5.33 PLHS502 block diagram.

of eight, one bank being D-types, the other bank R–S flip-flops. The clock for each flip-flop is derived from a separate small array which has direct inputs from four of the device input lines, four inputs from the main array and inputs from the Q and $\overline{Q}$ of each flip-flop. There is also an independent clear from the NAND array to each flip-flop. Both register banks are buried within the device, but with twenty-four output lines available there is an adequate capability to take the register contents out via the NAND array.

The performance of the devices has not been specified absolutely, as yet, but indications are that the delay times will be about 17 or 25 ns for one and two passes through the NAND array respectively. The supply current is expected to be about 250 mA and the devices will be supplied in a fifty-two-lead PLCC package. It has been suggested that these devices are equivalent to 3000 and 3600 two-input NAND gates respectively.

5.4.3 Programmable gate arrays

5.4.3.1 RAM-based programmability

All the devices described so far have been based on ROM technology, which means that once programmed they are unalterable without, in the case of MOS devices, going through an electrical or ultraviolet erase procedure. The function of the memory cell in a logic device is to define whether or not the input to a gate is open or closed. MOS PLDs use an MOS transistor for this; a HIGH voltage on the gate will allow the signal through while a LOW voltage opens the switch. A static RAM contains a number of storage cells which can be set to a HIGH or LOW state, then read at a later time if the power remains on the device. By connecting the output of each cell to the gate of a transistor in a logic array the stored level can be used to set the configuration of the logic array.

Figure 5.34 shows how similar the structure of a ROM-based PLD and RAM-based PLD can be. The disadvantage of the RAM-based PLD is that it loses its functionality when the power is removed, but many machines now use 'soft' program storage in place of the traditional PROMs. A RAM-based PLD can be loaded, along with the program RAMs, at switch on and hold its logic function

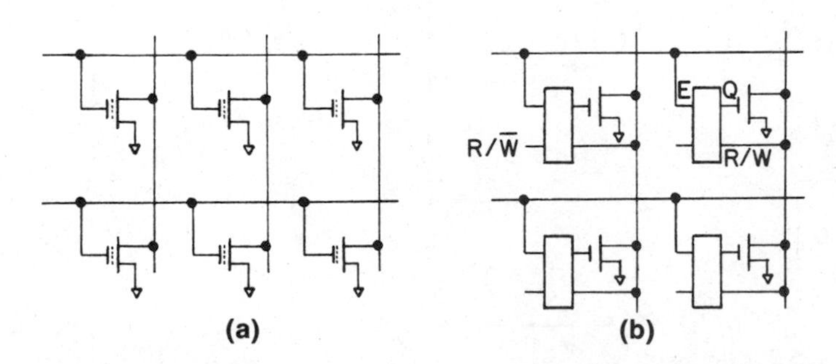

Fig. 5.34 (a) ROM based AND array; (b) RAM based AND array.

throughout the period of operation. The function of the PLD can even be changed while the system is running if the system function requires alteration. An advantage of RAM-based PLDs is that they do not need programming. This simplifies production and inventory management.

5.4.3.2 A proposed device

At present, samples of a device known as the XC–2064LCA have been produced. Its structure is more akin to a masked cell array than a PAL, as Figure 5.35 shows. It contains sixty-four logic cells, called *Configurable Logic Blocks* or CLBs, and fifty-eight I/O cells which are similar to the I/O cell of a 'V' PAL. The CLBs are separated by routing channels which carry the connecting links between them, and to the I/O cells. The connections between the routing channels and CLBs are defined by RAM cells, as are the configurations of the individual CLBs.

Each CLB contains a 'mini-PAL' and a flip-flop, as shown in Figure 5.36. The mini-PAL has four inputs from the routing channel and a feedback input from the flip-flop. Multiplexers define whether the flip-flop is driven from the routing channel or the mini-PAL, and whether the CLB output comes from the mini-PAL or flip-flop.

This device uses both a novel technology and novel architecture for PLDs, although both are well established in other devices. As with PML, it may take some time for users of traditional PLDs to come to terms with the different design techniques which will be needed to cope with a RAM-based gate array. Nevertheless, the trend in integrated circuit technology has always been towards more complex and highly populated structures, and these developments from the normal PLD architecture may be the way to achieve that goal.

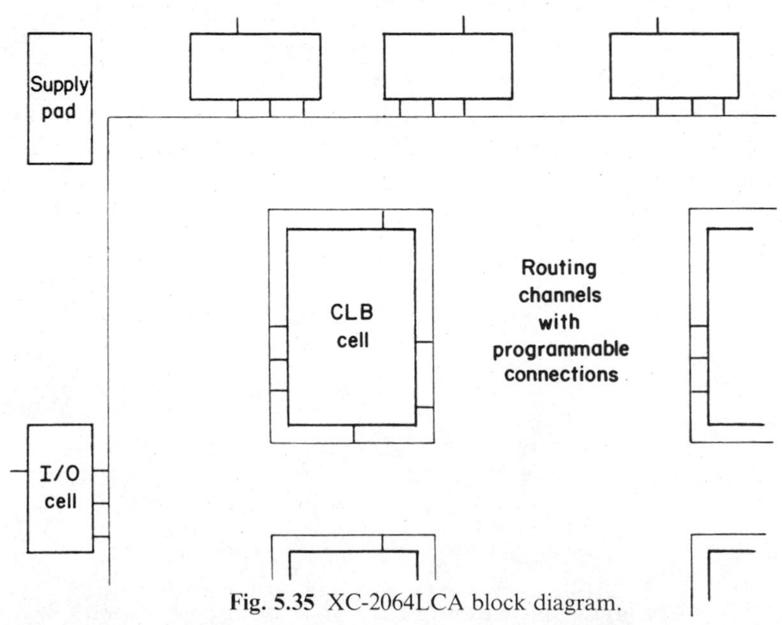

Fig. 5.35 XC-2064LCA block diagram.

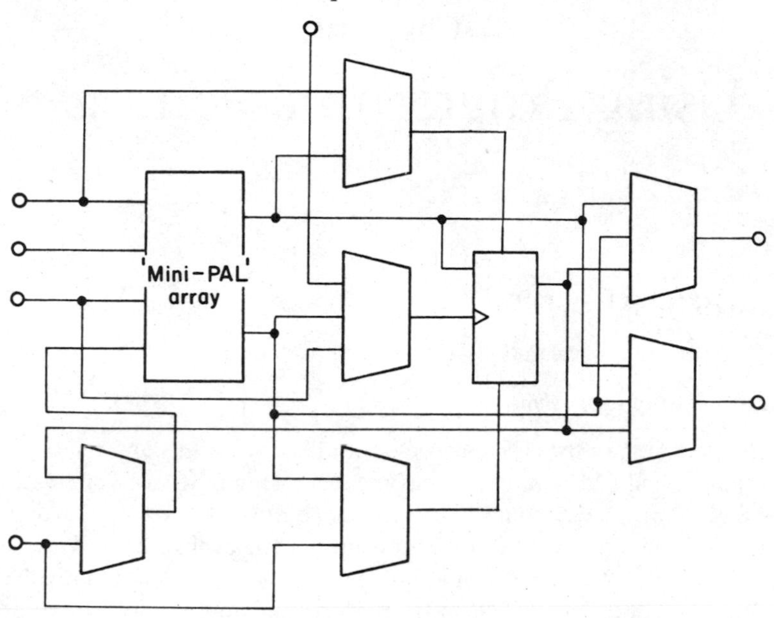

Fig. 5.36 Configurable logic block for RAM based PLD.

Chapter 6
Using Programmable Logic

6.1 WHEN TO USE PLDs

6.1.1 Economic considerations

6.1.1.1 Cost of ownership

Our investigation of the design process should start with a consideration of the criteria for using PLDs, rather than some other kind of device, for creating the logic circuit which we require. An early decision must involve the economic feasibility of this approach. To take the argument to an absurd limit, it would be possible to use a PLD in place of a simple gate, but to do so would involve four economic penalties: the PLD is more expensive, it would occupy more printed circuit board area, it would consume more power, and it would require programming. How, then, do we set about finding the right economic situation for using a PLD?

One way to quantify the factors involved in this decision is by using a principle known as *cost of ownership*. When we buy any commodity, the cost of owning it does not stop when we leave the saleroom. This applies to anything from a bag of flour to owning a house. Any householder knows that he has to pay taxes, repair bills, heating, decoration, furnishing, etc. in order to make a house habitable. Even converting a bag of flour to cakes involves the added costs of ingredients, cooking, washing-up and, not least, the cook's time. If we buy integrated circuits we do this, presumably, in order to build some more elaborate system which we then hope to sell. Each component in that system adds to the cost of that system by virtue of being part of that system. The various factors in that added value can be analysed to find the most economic way of building that system.

Before an integrated circuit can be used it has to be purchased, checked on arrival and stored ready for use. At some time before going to the production floor it may be tested; if it is a PLD it will need programming, either by the manufacturer or the user. It will probably be paid for before any cash is recovered by selling the system it is used in, so interest will be lost on its cash value. Some wastage will occur in handling the devices by accidental damage, using the wrong device or many other human failings. All this will happen before it gets near a printed circuit board.

The cost of the printed circuit board itself is directly related to the size of the integrated circuit and the number of pins it has. The board will need testing and some boards will fail. Using PLDs can have one of two consequences: the number of boards may be reduced, in which case handling costs will be reduced proportionately, or each board will be smaller, which makes the chance of a fault occurring less likely. In either case, there is a cost which can be associated with each component. There is also a labour cost associated with manufacturing and a 'hardware' cost, that is, the power supply, connectors and wiring loom as well as the box which the system is put in. Each component occupies space so more components means a bigger, and more expensive, box.

The final cost is concerned with problems after the system is delivered. Each component has a finite chance of failing, so each contributes to the expense of providing service under guarantee.

There are further costs which are less easy to quantify. Extensive use of PLDs shortens the development cycle, in itself a saving in engineers' salaries. The printed circuit board can be defined earlier, is easier to lay out and less likely to need changing. The product can be brought to the market earlier and, if it proves more reliable, will enhance the manufacturer's reputation for quality, thereby increasing sales. Thus, using PLDs may prove beneficial even where accountancy cannot prove an immediate cost saving.

6.1.1.2 Rentability calculation

One method of finding the cheapest solution is to perform a *rentability* calculation; in it one estimates the actual cost of each factor and adds them together for comparison. Several studies have been carried out in order to find the economic replacement value of PLDs (and other custom techniques). These all suffer from three assumptions which will be different in every case: what is being replaced, what is replacing it, and what the user's costs are. In the following analysis we can specify the first two exactly, but can only guess the third. However, each reader can put in the figures for his own situation and produce his own result. We will calculate three sets of costs: an LSTTL SSI circuit, a medium PLD (e.g. PLS153 or PAL16L8), and a complex PLD (e.g. PLS105 or PAL22V10). It is intended that the figures are in pounds sterling, but relative costs in the UK and USA are such that US dollars may be equally applicable:

Costing Category	*LSTTL*	*MCPLD*	*HCPLD*
Purchasing & stores	0.02	0.02	0.02
Testing & programming	0.05	0.12	0.14
Inventory & usage	0.03	0.10	0.18
PC board area	0.20	0.25	0.30
PC board fabrication	0.10	0.12	0.14
PC board test & rework	0.20	0.30	0.40
Power supplies & cooling	0.02	0.15	0.20
System hardware	0.05	0.055	0.06
Assembly labour	0.20	0.25	0.30
Servicing & reliability	0.07	0.07	0.07
Component cost	0.12	2.00	6.00
Total cost of ownership	1.06	3.435	7.81

We can, therefore, calculate how many LSTTL devices we need to replace to make it worth while using PLDs. Taking into account the other factors mentioned above it seems that a saving can be made by using a medium-complexity PLD in place of three SSI devices, or by using a high-complexity PLD in place of seven or eight. It must be emphasised that the above figures are estimates which will vary from application to application. However, the results are not untypical of the results obtained from most similar studies.

6.1.2 Technical considerations

6.1.2.1 Suitable application areas

If the economic analysis points towards PLDs the designer must still satisfy himself that the application is suitable. This is a technical decision. PLDs are not restricted to any particular sphere of application; every designer of electronic circuits should count them as a potential solution to his design problems. The alternatives to PLDs are discrete logic or masked ASICs and these, too, are universal in their application areas. The only reasons for not using PLDs are economic, or because they cannot provide an adequate performance.

6.1.2.2 Performance limitations

The technical limitations of PLDs are likely to be concerned with speed, power consumption or simply fitting the circuit into a single package.

The fastest PLDs compare well with the fastest discrete logic circuits. On the face of it they may look slower – 15 ns delay compared with 5 ns or less for the fastest TTL-type devices. However, PLDs may replace two or more levels of logic, so the effect on the system will be the same as the discrete logic solution. Part of the reason for this multi-level replaceability is that the structure is based on an AND–OR architecture. Another factor is that PLDs use very wide gates, up to 20 inputs; to build a 20-input gate from discrete logic takes two levels of logic, hence the saving.

Now that CMOS is becoming more widely used for PLDs, there are fewer problems arising from high power consumption. A typical gate package from a TTL family consumes less than 10 mA, while a low-complexity bipolar PLD takes about 100 mA. A replacement factor of more than 10 is needed to give a comparable power. Some designers will accept this as the price to pay for the convenience of using PLDs, but any reduction in power gives benefits to the user. The majority of CMOS circuits can offer speed equivalent to standard bipolar PLDs, with about half the power, by building the peripheral components from true CMOS. Circuits which are truly CMOS are much slower than bipolar, but have supply currents measured in microamps. It should be remembered that CMOS power dissipation increases with frequency, so the benefit reduces as the operating frequency increases.

With new structures being announced regularly, part of the designer's problem is finding the best device to fit his circuit. The major constraint is the number of pins available in the package for I/O. If this is not enough for the circuit under consideration, more than one device may be necessary. In this case, the number of inputs may be the critical factor for each output may need most of the inputs, which must then be connected to each package. The exception is registered PLDs for state diagrams, when the outputs are fed back to the array input.

Figure 6.1 is a table listing all the PLDs currently available. They are sorted into columns with the same number of inputs and rows by complexity; the table may be used to select the best device for any application. Because input count is probably the most important factor, initial selection should be based on this. The column with the required number of inputs will contain appropriate devices in order of complexity. If the least complex device does not have enough AND terms or outputs, the next device may be considered, and so on. Alternatively, a device with more inputs can be considered, bearing in mind that in many cases inputs can be traded for outputs, thanks to bidirectional I/O pins.

	5/6	8	9	10	11	12	13	14	16	18	20	22	32	36	38	64
LSI									32R16		EP600 5C060		PLHS502 64R32 PLHS501	EP1200 5C121	EP900	EP1800
FPLSs		PLS159		PLS157		PLS168 PLS155		PLS167	PLS105 PLS169							
REGd PLEs			9R8	10R8	11RA8 11RS8											
V PALs									16V8	EP300		22V10				
COMPLEX REG PALs				20RS10 20X10		20RS8 20X8 16A4 16X4			20R64 20X4		20RA10					
REGd PALs		16R8		16R6		20R8 16R4		20R6	20R4							
8 BIT PLEs	5P8	8P8	9P8	10P8	11P8	12P8	13P8									
4 BIT PLEs		8P4	9P4	10P4	11P4	12P4										
FPLAs						PLS161			PLS100	PLS153		PLS173				
COMP PALs										18P8	20S10					
MC PALs									16*8		20L8 20L10					
LC PALs				10*8		12L10 12*6		14L8 14*4	16L6 16*2 16C1	18L4	20L2 20C1					
FPADs	6L16	8L14				PLS163			PLS103 PLS162	PLS151						

* - 'H', 'L' or 'P'

Fig. 6.1 Summary of PLDs available or announced.

6.2 METHODICAL APPROACH TO PLD DESIGN

6.2.1 Outline of the steps

6.2.1.1 How to start!

There is an apocryphal story of the local resident who, on being asked directions, stated, 'If I were going there I wouldn't start from here.' So, how does one go about setting out on the path to a logic circuit?

Most systems take an input from the outside world, manipulate it in some way and then present the result to the outside world again. If the manipulation is carried out by a logic circuit then the information must be converted to a form which the logic can relate to, and then converted back to a form in which the outside can understand it. Examples are: computers, instrumentation, communications circuits and even PLD programmers. The conversions are the job of interfaces, the manipulations are done by the logic circuits. In many cases the main job of the logic can be undertaken by microprocessors or dedicated LSI circuits, which frequently need combinational logic to glue them together. Sometimes the whole logic function may be implemented by a collection of standard logic circuits.

In either case there is likely to be an area where PLDs may be used to advantage. Combinational PLDs more often fall into the glue category, while registered PLDs are frequently used to make broader-based functions. These boundaries are by no means rigid and both lend themselves to the same basic design techniques. In most cases, though, there will be some dedicated LSI involved so the design will start by defining these components.

The next step is to decide how to connect the LSIs together, and generally some help is forthcoming from their data sheets to indicate which signals they need to work together. At this stage it is usually necessary to start inserting inverters, gates and more complex functions into signal paths. This is because control signals from one device are not exactly what is required by another, particularly if they come from different families of LSI. The end result is a circuit diagram containing a few LSI devices connected by several discrete logic devices.

6.2.1.2 Partitioning the circuit

If we have a circuit diagram, the design has only just started. We have to decide which physical devices are best suited to actually implementing the logic function we have drawn. There will be constraints on this decision:

- how much time can we allow for signals to pass between devices?
- how big can we make the printed circuit board?
- how much power can we afford to dissipate?

Let us assume that the answers do not rule out PLDs, they may even make the use of PLDs mandatory. The first stage in designing the PLDs which we are proposing to use is a process called *circuit partitioning*. This is done by drawing a box round the parts of the circuit containing logic functions which are not incorporated into the LSI; we can now count the number of signals entering the box, the number of signals leaving the box and the number of gates or equivalent functions contained inside the box. The lower the sum of the first two numbers and the higher the third number, the more likely it is that a PLD would be the best solution. The chances are that with less than twenty-five inputs and outputs and more than twelve gates, it will be worth considering a PLD.

The justification for this is that twenty-four is the highest number of I/Os in a low- or medium-complexity PLD, and more than twelve gates probably means more than three discrete logic circuits, which we have seen is the minimum number worth replacing. If the number of I/Os is substantially higher than twenty-four it may be worth considering using more than one PLD. In this case the partitioning needs to be done to make several boxes, all of which conform to the criteria above. If the number of gates is too low to make two or more PLDs economic then one solution might be to use a PLD to incorporate most of the logic, together with discrete logic to take care of the left-overs.

6.2.1.3 The design steps

We have reached the stage where we have one or more blocks of logic defined by logic symbols connected by signal paths. The next step is to convert this into a format suitable for entering this data into a PLD programmer. The most common way to define PLDs is by Boolean logic equations, so we should convert the logic diagram to equations. Starting from the inputs we can give each signal path in turn its logic equivalent until we arrive at the outputs. In this way we have defined each output as a logic combination of the input signals. Because PLDs are based on an AND–OR structure the equations have to be converted to the same format, which is often called *sum of products*. This is because the AND operator is manipulated in a similar way to multiplication, and the OR operator similarly to addition.

Conversion of the equations may be done manually although it is more usual to use a computer program; some software will also handle the logic diagram directly. At this point we can see how many AND gates are required for each of the outputs and an initial selection of PLD can be made. The factors which will influence this choice are:

- number of inputs
- number of outputs
- number of output signals fed back as inputs
- number of AND gates required for each output
- maximum delay time allowable
- maximum power consumption allowable

The procedure suggested in the previous section should indicate the preferred choice. If this does not give a simple PLD as the outcome, there are some steps which can be taken to try to simplify the logic. Firstly, it may be worth trying a standard logic reduction technique, such as Karnaugh mapping described earlier in this book. This may reduce the number of gates required sufficiently to enable the use of a low-complexity device and, hence, reduce cost.

A second consideration is the use of bidirectional pins. These may appear to be necessary if output signals are fed back as inputs, but it may be possible to generate the 'secondary' logic function directly from the inputs which will confer two benefits to the design. It will remove the need for bidirectional pins and therefore the need to use medium-complexity PALs, provided that low-complexity devices have enough AND gates. Also, there will be an improvement in delay time as the signals will not have to make two passes through the PLD.

Lastly, if output signals have to be fed back but the gate count is quite low, it would be worth investigating the possibility of feeding back externally if there are any spare input pins. This again could make it possible to use a low-complexity device in place of a more expensive part.

6.2.1.4 Completing the design

Having reached the stage of possessing a minimised set of logic equations, and deciding on the target device, all that remains is to complete the design. This means that the logic information has to be mapped on to the fuse chart of the PLD. The method of doing this depends on the tools available to the designer. Software tools will be described in a later chapter, but whichever method is used the information will have to be entered into either a computer or a programmer.

PLAs may usually be entered directly as a truth table so, without software, the problem facing the designer is to convert logic equations into a truth table. We addressed this problem in Section 4.2.3.1, but we may summarise the conclusion here. If the input variable appears in the equation without inversion then enter an 'H' in the truth table, if it appears inverted enter an 'L'; otherwise, if it does not appear at all enter '–'. On the output side, if the output includes the equation enter 'A'; otherwise, enter '.'. If a bidirectional pin is being dedicated as an output, all the inputs should contain '–'s; if it is an input all inputs should be entered as '0'; otherwise, the enabling equation should be entered for true tri-state operation.

The last question to be settled is, will the design perform the function which is intended? This can usually be solved by defining *test vectors* which are used to simulate the device performance, but need a computer program to make them operate. If this step is successful, the test vectors can be used to test real devices functionally, as we shall see later in this chapter.

6.3 PROGRAMMING

6.3.1 PAL/PLA fuse mapping

6.3.1.1 Programming map

We saw, in Chapter 2, how PROMs use the same addressing circuitry to enable the fuses for both programming and reading. This is because the input lines are fully decoded in the AND array. PALs and PLAs have a programmable AND array so the input lines cannot be used for decoding the fuses directly. Additional circuitry is needed to address the fuses and this is enabled by applying supervoltages to specified pins. The inputs pins then become equivalent to PROM inputs and address the fuse array as if it were a PROM. By applying supervoltages to the supply and outputs the programming circuitry itself is enabled.

Because the fuses are programmed as a PROM they are given a pseudo-PROM structure for programming purposes. This address map only refers to the programming situation and need not concern the design situation as it should be taken care of by the programming equipment. In practice, PALs in the same package size all use the same addressing method. This means that for smaller PALs it is possible to address fuses which do not really exist. These are called *phantom fuses* and are usually shown in X-charts as a '.'; they are functionally equivalent to a blown fuse, '–'.

6.3.1.2 JEDEC fuse map

The other way of specifying fuse location is by the JEDEC fuse map. JEDEC is the Joint Electronic Devices Engineering Council; it has the task of ensuring that specifications for the same device from different manufacturers are compatible. This makes certain that a design completed on one manufacturer's device can be transferred to another manufacturer without changing the design parameters. It also gives producers of design software a standard format for loading their output into programming equipment.

JEDEC fuse numbers are not the same as the fuse address, which is the binary code used by the programmer. They run in sequence from '1' to the total number of fuses within the device and, usually, follow the order which is specified by the internal fuse map but omit the phantom fuses. A full JEDEC file contains a full description of the device being programmed, apart from a list of blown fuses. A typical JEDEC file might appear as below:

```
STX (i.e. control + B)
P.L. Systems Inc                                    (Company name)
A.C. Guy                                            (Designer)
30 06 86                                            (Date)
ABC9876                                             (Drawing number)
PAL16L8                                             (Part number)
*
*D2029                                              (Device code)
*F0                                                 (Default fuse state)
*G0                                                 (Security fuse state)
*L0000 11111111111111111111111111111111             (Fuse information)
*L0032 1010111011101111001011110100111              (Fuse information)
*L0256 11111111111111111111111111111111             (Fuse information)
.

.
*L1920 11101110101001110110111011010101             (Fuse information)
*C4AB7                                              (Fuse checksum)
*V0001 100111000N1HLLLHHXXN                         (Test vector)
*V0002 110011011N0LLLHXXLHN                         (Test vector)
.

.
*V0100 011100101N0HHHHLLLLN                         (Test vector)
*
ETX (i.e. control +C)
B642                                                (Transmission)
Checksum
```

Commercial programmers will accept data in the JEDEC format and translate it into a fuse map corresponding to the device structure in programming mode.

6.3.2 Device programmability

6.3.2.1 Intrinsic yield

The one part of a PLD which cannot be checked by manufacturers using metal or diode fuses is the programming circuit; any attempt to test it will cause part of the device to be programmed and therefore rendered useless. On the other hand, UV or electrically erasable cells can be programmed and then erased so it is possible to check that these devices are programmable before releasing them for delivery. The intrinsic yield of these devices, which tend to be MOS or CMOS, is likely to be higher than that of bipolar PLDs.

As with bipolar PROMs, the manufacturers of bipolar devices usually include extra fuses, which play no part in normal device operation, to test programmability. This, again, has the spin-off of allowing testing of the programmed output level which could not be guaranteed otherwise. Other checks to ensure good programmability are carried out during the manufacturing process. These include resistivity and thickness of the deposited metal layers, width of the fuses after etching and visual inspection of the circuits for damage to diffusion and metal layers after etching. Advances in manufacturing techniques such as ion implantation, plasma etching and electron beam lithography also reduce defect densities and contribute to higher programming yields.

6.3.2.2 Batch effects

Batch effects may occur due to manufacturing defects or to statistical fluctuations in the received devices. Considering manufacturing effects first, there are several areas where a defect might cause a batch problem with programmability. This starts at the design stage where a single mistake could cause a device to be over-sensitive to one of the process parameters; modern *Computer Aided Engineering*, or CAE, techniques make this event unlikely. A single mask fault might cause the circuit in the same position on every wafer to be defective, while an out-of-tolerance etching or layer deposition stage could cause a whole batch to be marginal. The effect of this on the overall programming yield depends on the number of defective circuits per batch and the number of bad batches in the production. It is the task of the manufacturer's quality control department to prevent these faults reaching the user. Bad batches will tend to get diluted by good product in the subsequent stages of assembly, testing and distribution and appear as an overall programming yield to the user.

The actual number of rejects seen by the user depends on the overall proportion of defective devices and statistical fluctuations. For example, if the overall rate is 1 per cent, then normally no rejects would be seen if just ten devices were programmed, but just occasionally – about nine times in every hundred –one reject will occur. Less often, once out of two hundred times, two rejects will be found in ten which can give a misleading impression of the true rate on the odd occasion when it happens. The larger the batch, the more likely is the observed failure rate to be close to the true rate, as the following table shows:

TRUE RATE – 2%	*Observed Rate*						
Batch Size	*0%*	*1%*	*2%*	*3%*	*4%*	*5%*	*6%*
100	0.13	0.27	0.27	0.18	0.09	0.04	0.01
300	0.02	0.26	0.46	0.21	0.04	0.01	
1000		0.06	0.73	0.20	0.01		
3000		0.02	0.94	0.04			

The figure in each column is the probability of observing a failure rate centred on the value at the head of the column. Thus, in a batch of a hundred devices, there is a probability of 0.18 of finding three rejects; in a batch of a thousand there is a probability of 0.06 of finding from five to fourteen rejects. As might be expected, there is a lower chance of finding an excessive reject rate in larger batches, although the possibility of statistical fluctuations must be considered if a high reject rate is found in a batch of PLDs.

6.3.2.3 User problems

If devices are not programmed properly then, however good the quality of the product, the yield will be poor. In this section we examine the problems which a user can bring on himself because he is using poorly maintained equipment for programming. The most important factor is to ensure that the equipment meets the specification laid down by the device manufacturer. Every aspect of the specification is normally included to an exact tolerance and if the tolerance is exceeded there can be a damaging effect on the yield. The parameters specified will include pin voltages, current capability of the power supplies, width and edge speeds of pulses, relative timings of pulses, average power consumption and allowable number of programming attempts.

If the tolerances are exceeded, the programming yield may suffer, as we saw above; it is worse, though, if the device is programmed in an unreliable manner. It is worse to use unreliable devices than to waste PLDs needlessly through misprogramming. The latter costs the price of the device while the former may cause equipment to fail in the field, which involves repair costs, and may result in lost reputation. If yields are lower than expected, the programming equipment should be calibrated to eliminate the possibility of drift in the equipment.

If the electrical parameters are all in specification it is still possible for the equipment to be causing extra failures. The other critical part of the programmer is the socket. If contact between the socket and device pins is poor there can be two ways in which performance is likely to be affected. As we have seen before, a relatively high current is needed to blow the fuses in bipolar PLDs. A high resistance between socket and device pins can cause the voltage to drop below the level needed to turn on the programming circuit in the PLD. The most likely result is to cause a 'no blow' when the device fails to program but is not otherwise damaged.

If there is poor contact on one of the addressing pins it is possible that the device will 'see' a HIGH although the programmer has put a LOW on to the socket pin. This will cause the wrong fuse to be programmed and may result in the device failing to verify after the programming sequence. If a HIGH is misread every time a LOW is presented then half the fuses will be programmed at the wrong time; the chance that this will not result in failure is negligible.

Most programmers use *zero insertion force* sockets, which allow the device to be dropped in and then contact is made by clamping the pins in the socket. These sockets can age by the cam wearing so that the clamping is ineffective, or by the plating on the contacts wearing or becoming contaminated. Regular cleaning is an obvious remedy; in addition, the sockets should be changed regularly before they start to wear out. After all, if a socket costing £10 (or $14) is changed when 10 000 devices have been programmed, the cost is only 0.1 p (1.4 c) per device. This is paid for by a yield improvement of 0.05 per cent on a £2 ($2.8) device.

6.4 TESTING

6.4.1 Why test?

6.4.1.1 Potential PLD faults

We noted in the previous section that PALs and PLAs need a separate fusing circuit in order to address the fuse array as a pseudo-PROM. A programmer will usually follow a standard sequence during programming, as we saw before. The final stage in the sequence is a verification that the correct fuses have been blown, but the fuse addressing circuit has to be used for this. Faults in this circuit will be discovered in the verification but the possibility remains that there are faults in the 'logic part' of the PLD. This argument does not apply to PROMs as these use the same circuit for fuse addressing during programming and during normal operation.

The effect of this is not hard to imagine; having programmed a batch of PLDs and inserted them into his circuit, the user will find that some of the circuits do not work as he expected. This is because the PLDs are not following the correct logic pattern even though all the fuses appear to be correctly blown. This could be due to a fault in the fuse addressing circuit, but it is more likely to be because there is a defect in the logic path which has not been detected by the fuse verification. The solution is to test the logic path independently.

6.4.1.2 Test classes

Testing is a global concept when applied to any product and to understand the problems we need to define test classes. As far as this discussion goes we are interested only in *zero-hour tests*. These, as their name implies, refer to the performance of devices when they are unused. *Life tests*, on the other hand, measure how reliable devices are by seeing how long it takes them to fail, failure usually being measured as an inability to pass the zero-hour tests.

Zero-hour tests themselves can be divided into two groups, *parametric* and *functional* tests. Parametric tests usually measure the ability of the device to interface to other parts of the system, covering such features as the current taken by the inputs, current drive available from the outputs, voltage levels defining HIGH and LOW logic states and power consumption. These are often called d.c. tests. Also included in parametric tests are a.c. tests which measure timings relevant to circuit operation, such as propagation delays, set-up and hold times and clocking speeds.

Parametric test results are not usually affected to a large extent by programming. There may be a small reduction in power consumption if some AND terms are removed, and this may also cause some speed improvement. On the whole though, any parameters measured and guaranteed by the manufacturer should not need further testing. We noted before that the test row and column are used to set the device to its active output condition; this allows the manufacturer to carry out full parametric testing.

Functional testing is another matter. An unprogrammed PLD will be completely inactive, apart from the test fuses, so there is no way of knowing whether the logic path is working fully. Whatever pattern of inputs is applied, the outputs will not change, so there is no way of applying a functional test until after programming. A functional test, at its simplest, consists of applying different patterns of HIGHs and LOWs to the inputs, and checking that the correct pattern is obtained from the outputs. The input patterns are called *test vectors* and the skill in designing functional tests is to create an optimum set of test vectors.

6.4.2 Test coverage and testability

6.4.2.1 Design for test

The first criterion in designing a test program is whether the circuit to be tested is testable or not. At first sight it might appear that any circuit must be testable, but this is not necessarily the case. Consider the simple D-latch; the simplest equation is:

$$
\begin{aligned}
Q = {} & D * LE \\
& + Q * \overline{LE}
\end{aligned}
$$

This is testable, provided that we start with LE HIGH, because a fault in any of the three gates can be detected. Every combination of HIGHs and LOWs can be applied to each gate and the resulting output verified without interference from any other gate. However, to achieve a glitch-free output, we modified the design to:

$$
\begin{aligned}
Q = {} & D * LE \\
& + Q * \overline{LE} \\
& + D * Q
\end{aligned}
$$

Now we find that some of the gates interfere with each other. In particular, we cannot verify the action of the new AND gate which we have added, not surprisingly as it does not alter the function of the circuit. In order to make the circuit fully testable we would have to add a test control signal 'T' which could isolate the third gate. The testable glitch-free circuit becomes:

$$
\begin{aligned}
Q = & D * LE * T \\
& + Q * \overline{LE} * T \\
& + D * Q
\end{aligned}
$$

Taking 'T' LOW will disable the first two gates and enable the third gate to be tested in isolation, having first set Q to a known level. This is an example of a circuit being untestable because of parallel logic paths; this can be resolved by disabling one path for the test sequence.

A second cause of untestability is starting in an unknown state. For example, a simple counter will increment its count by 1 for each clock pulse it receives, so part of the test sequence will be to clock the counter and check for an increase of 1. If we do not know what the first count state is, we do not know which state to check for after the clock pulse. The remedy is equally obvious; provide a means of setting the counter to a known state, if necessary by supplying an additional reset signal for test purposes only.

A similar problem is that of illegal states. A decade counter, for instance, will not normally use output combinations corresponding to the hexadecimal numbers A–F, but it may be possible for it to enter those states accidentally; a parallel load function might allow these numbers to be entered. If no transition path is available out of these states the counter will be stuck there indefinitely. It is therefore advisable to provide a state jump from all 'illegal' states to a known state. The complement term is a convenient method of doing this in those devices which possess that feature.

The fourth cause of untestability which we shall examine is the case where there is a dead-end in the state diagram. It is difficult to envisage this being designed-in deliberately, but it could occur accidentally by a poor design allowing a particular input condition to trigger two jumps at once. If there is an inadvertent dead-end then, again, once entered the device will become 'stuck'. In particular, care should be taken when using the 'hold' type of feedback with D-register PLDs, for a hold is a dead-end unless some way of releasing it is included. The solution, as in the previous case, is to provide a default jump out of illegal states.

6.4.2.2 Fault grading

We have discussed the need for testing to find faults and we will now proceed to describe in more detail how to define them, to ensure that the test routines will check every possible fault. If we look at the simplest logic device, the inverter, we can see how faults may be described. When functioning correctly, the output from an inverter will be the opposite sense from the input; a faulty inverter will have an output which is either always HIGH or always LOW. These faults are referred to as *stuck-at-one* or *stuck-at-zero*, SA1 or SA0 for short. There is a third possibility, that the output is the same as the input. This condition is most unlikely as most faults are due to open-circuits or short-circuits to one of the supply rails; fault analysis is much easier to handle if only SA1 and SA0 faults are considered.

Fault grading is the process of finding how many of the possible SA1 and SA0 faults are detected by the test vectors being graded. Applying a HIGH to the input of an inverter will detect an SA1 output, for if the output is SA1 the test will fail. To detect an SA0 output as well it is necessary to apply a LOW to the input, for then we expect the output to go HIGH. We can illustrate this by looking at the old example of the simple combination lock. This had the equation:

$$\text{OPEN} = A * \overline{B} * \overline{C} * D$$
$$+ A * \overline{B} * C * \overline{D}$$

Figure 6.2 shows how this would be programmed in a simple PLA. It shows that there are nine possible nodes in the circuit, each of which must be tested for SA1 and SA0 faults. The test vector A–H:B–L:C–L:D–H tests lines 1,2,4,5,7,9 for SA0, for if any of these is stuck LOW the output will be LOW instead of HIGH. We can write out a full table of vectors to cover all possible faults in this circuit:

A	*B*	*C*	*D*	*OPEN*	
H	L	L	H	HIGH	SA0–1,2,4,5,7,9
H	L	H	L	HIGH	SA0–1,2,3,6,8,9
L	L	L	H	LOW	SA1–1,7,8,9
H	H	H	L	LOW	SA1–2,7,8,9
H	L	H	H	LOW	SA1–4,6,7,8,9
H	L	L	L	LOW	SA1–3,5,7,8,9

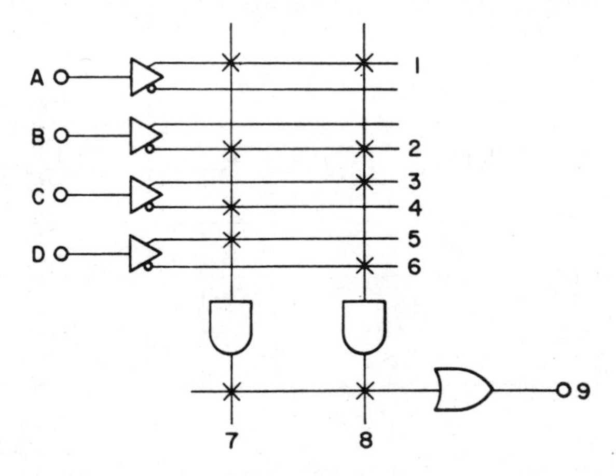

Fig. 6.2 PLA programmed with simple combination lock.

Note that there are sixteen possible test vector combinations from the four inputs, but that we have covered all possible faults with just six of these. This is an important fact because it means that a circuit with n inputs may be fully tested with many fewer than the 2^n possible combinations of test vector. In PLD circuits n can be as high as 20 which gives over a million possible combinations.

6.4.3 Designing test sequences

6.4.3.1 Simple combinational circuits

We can use the result from Section 6.4.2.2 to see the principles behind designing test sequences. To start with we can look at the simplest circuits, that is, those using just combinational logic without feedback. Once again we will use the Karnaugh map as a tool; this time it will help us define the test vectors. Figure 6.3 shows the map for the logic function of the combination lock and a map on which the test vectors are plotted. The first two vectors correspond to the logic function and, in effect, check that the circuit responds to its active logic inputs. In other words, it ensures that the output goes HIGH when it should; these are sometimes called type-1 tests.

The other four vectors test the condition that the output stays LOW when the inputs are not in an active combination, that is, that the output does not go HIGH when it should not, often referred to as type-2 tests. The Karnaugh maps show that the type-2 vectors are all adjacent to the type-1 vectors. This ensures that a fault on one input will not be masked by other inputs. If the starred cell was used as a vector and line 1 (Figure 6.2) was SA1, the test would still pass because input B is in an inactive state. The vector would not detect SA1 faults for either line 1 or line 2 unless, of course, both lines were faulty.

The Karnaugh map is useful in the mechanical process of defining output levels corresponding to input vectors in more complex systems. Most circuits will have several outputs and each output must be fully tested for SA1 and SA0 faults; often one input vector will be relevant to more than one output and more

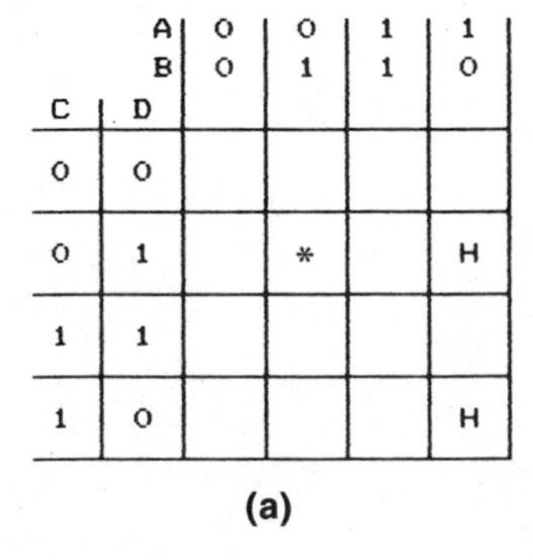

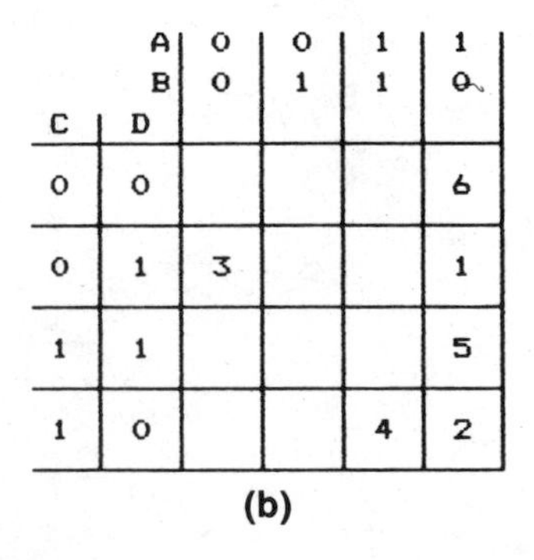

Fig. 6.3 Karnaugh maps. (a) combinational lock; (b) test vectors.

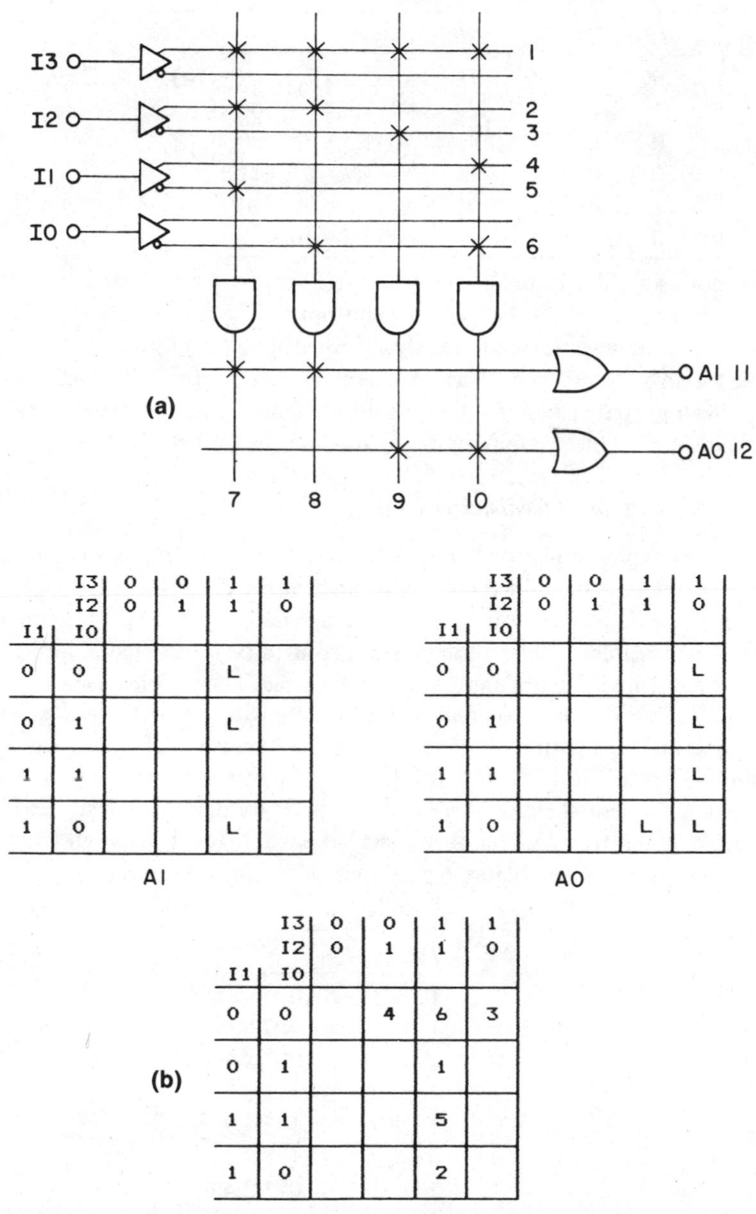

Fig. 6.4 4-input priority encoder. (a) PLA circuit; (b) Karnaugh maps.

coverage will be obtained if every output is checked for each input combination. As an example of a more complex circuit we will derive the test vectors for a four-input priority encoder. The PLA circuit and Karnaugh maps are shown in Figure 6.4. Input vectors are conventionally indicated by '1'(HIGH) and '0'(LOW), while the resulting outputs are given 'H' and 'L'. The table of test vectors is:

I3	*I2*	*I1*	*I0*	*A1*	*A0*	
1	1	0	1	L	H	SA0–1,2,5,7,11 SA1–3,9,10,12
1	1	1	0	L	L	SA0–1,2,4,6,8,10,11,12
1	0	0	0	H	L	SA0–1,3,9,12 SA1–2,7,8,11
0	1	0	0	H	H	SA1–1,7,8,9,10,11,12
1	1	1	1	H	H	SA1–3,5,6,7,8,9,10,11,12
1	1	0	0	L	H	SA1–3,4,9,10,12

Various points are illustrated by this example. Vectors 1 and 3 can cover both a type-1 and a type-2 test for the respective outputs; every input is defined for each vector, even though some inputs are 'don't care' for some outputs; vector 6 is not suitable for an SA0 test on A1 because this is an overlapping cell which falls into the parallel logic path category of untestability. Once again, we need only six out of the sixteen possible combinations to cover all the faults.

6.4.3.2 *Combinational feedback circuits*

The same principles apply to circuits with feedback as to simple combinational circuits, but there is the added complication that the level of the fedback input may need to be defined by other inputs. An alternative approach is to break the feedback, if possible, by using the tri-state control or a test enable input. This may not be possible if 'spare' inputs are unavailable. The problem is not so acute when the feedback is used to create logic circuits with several levels, as with a parity generator which may have a number of exclusive-OR gates cascaded. These may be tested as in Section 6.4.3.1, but circuits such as the D-latch must be treated as sequential circuits. We can see this if we derive the test vectors for the testable 'glitch-free' D-latch from Section 6.4.2.2. The Karnaugh maps and PLA circuit are shown in Figure 6.5 and yield the following vectors:

D	*LE*	*T*	*Q*	
1	1	1	H	SA0–9
1	0	1	H	SA0–9
0	0	1	H	SA0–3,4,5,7,9
0	1	1	L	SA1–1,6,9
0	0	1	L	SA1–5,7,9
1	0	0	L	SA1–1,8,9
1	1	1	H	SA0–1,2,4,6,9
1	1	0	H	SA0–1,5,8,9
0	1	0	L	SA1–4,6,9

Because of the feedback we are unable to isolate LE so that lines 2 and 3 cannot be tested for SA1 faults. It may be seen from the Karnaugh map that the tests progress in sequence round the map one input at a time. The arrows show where the output sense changes with a consequent change in map position, again because of the feedback. By changing only one input at a time we avoid timing problems. If we took LE LOW and changed the sense of D in the same test, then

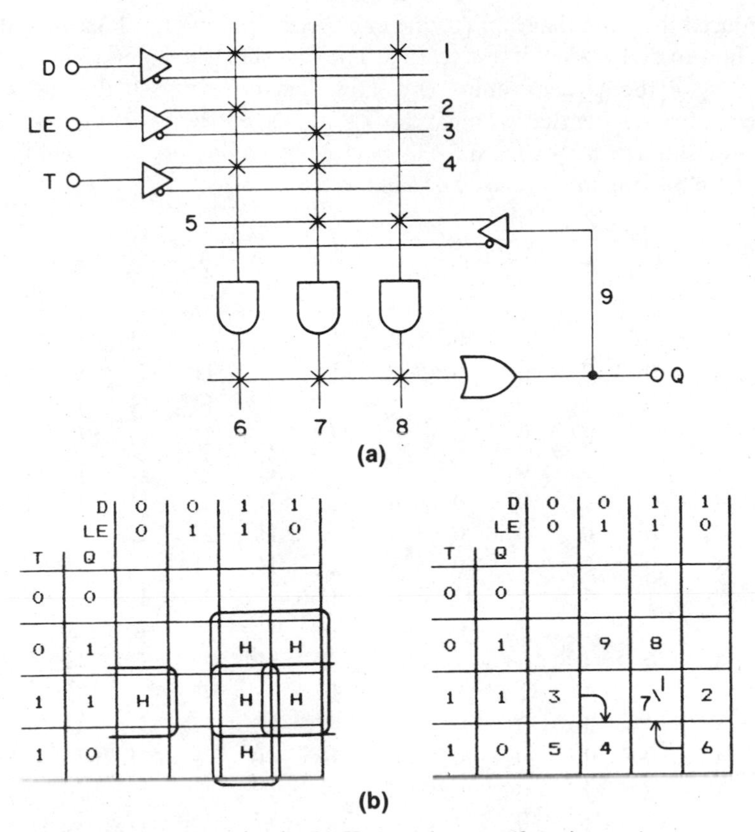

Fig. 6.5 D-latch. (a) PLA circuit; (b) Karnaugh maps (function and test vectors).

the output level would depend on which change occurred first. This race condition should be avoided in designing test sequences as much as in designing the circuits themselves.

6.4.3.3 Sequential circuits

We will assume that by sequential circuits we mean state machines. Test sequences for state machines can become very complex; usually it is best to consider transitions from each state in turn, in which case it may be possible to treat each state in a similar fashion to a combinational circuit. Type-1 tests are then those which are designed to cause a state jump, while type-2 tests are those which should leave the state register unchanged. The main complication occurs with devices containing buried registers, although in some cases it is possible to read these via output pins by applying a supervoltage to a specified input.

Rather than base the test sequence on an analysis of SA0 and SA1 faults of the internal components it is usually simpler to use the state diagram as a basis. In practice the result will be almost the same and will have the advantage that it will probably mirror the final application more exactly. In Figure 6.6, we have reproduced the state diagram for the enhanced combination lock in order to show how to derive a set of test vectors. The first step is to enter a known state, by means of the asynchronous reset, and then run through the states with 'correct' entries. After that we must check that false entries take us into the error states; we still use the SA0 and SA1 principle by having only one bit at a time false when moving into errors. The full table is:

CK	*RS*	*I3*	*I2*	*I1*	*I0*	*Q3*	*Q2*	*Q1*	*Q0*	*UN*
0	1	1	1	1	1	L	L	L	L	L
C	0	1	0	0	0	L	L	L	H	L
C	0	0	0	0	0	L	L	H	L	L
C	0	1	0	0	1	L	L	H	H	H
0	1	0	0	0	0	L	L	L	L	L
C	0	0	0	0	0	L	H	L	L	L
C	0	1	0	0	0	L	H	L	H	L
C	0	0	0	0	0	L	H	H	L	L
C	0	1	0	0	1	L	H	H	H	H
0	1	1	1	1	0	L	L	L	L	L
C	0	1	1	0	0	L	H	L	L	L
C	0	1	0	1	0	H	H	L	L	L
0	1	1	1	0	1	L	L	L	L	L
C	0	1	0	0	1	L	H	L	L	L
C	0	1	0	0	0	L	H	L	H	L
C	0	1	0	0	0	H	H	L	H	L
0	1	1	0	1	1	L	L	L	L	L
C	0	1	0	0	0	L	L	L	H	L
C	0	0	1	0	0	L	H	L	H	L
C	0	0	0	1	0	H	H	L	H	L
0	1	0	1	1	1	L	L	L	L	L
C	0	1	0	0	0	L	L	L	H	L
C	0	0	0	0	1	L	H	L	H	L
C	0	0	0	0	0	L	H	H	L	L
C	0	1	0	0	0	H	H	H	L	L
0	1	1	1	1	0	L	L	L	L	L
C	0	1	0	0	0	L	L	L	H	L
C	0	0	0	0	0	L	L	H	L	L
C	0	1	0	1	1	L	H	H	L	L
C	0	1	1	0	1	H	H	H	L	L
0	1	0	1	1	1	L	L	L	L	L
C	0	1	0	0	0	L	L	L	H	L
C	0	0	0	0	0	L	L	H	L	L
C	0	0	0	0	1	L	H	H	L	L

The full test sequence thus takes thirty-four vectors. One reason for this is that every state has to be tested individually as a simple combinational circuit, so the number of tests per circuit has to be multiplied by the number of states to find

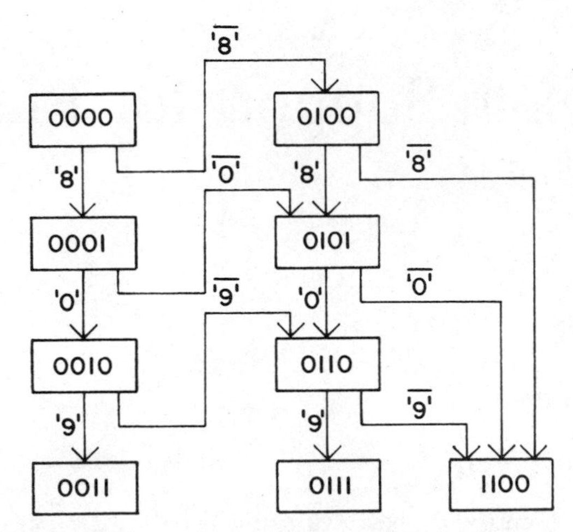

Fig. 6.6 Enhanced combination lock state diagram.

the total number of tests. The other reason is that once a state has changed it must be re-entered before it can be tested again. Part of the above sequence is taken up by moving round the state diagram back to test the appropriate state with another input. Some PLDs have the ability to load the output register by means of a preload function or input supervoltage; the need for such a facility is amply demonstrated by the test sequence we have just created.

6.4.4 Summary of test design procedure

6.4.4.1 Pitfalls

Testing is such an important stage in designing PLDs that we make no apologies for summarising the above discussion. Pitfalls to be avoided as part of design for test and test design are:

- parallel logic paths
- undefined entry states
- dead-ends (states with no exit)
- race conditions where the output depends on the order in which signals are applied
- floating inputs (unless they are not used by *any* AND-term)

6.4.4.2 Procedures

The following comments can apply to either combinational or sequential circuits bearing in mind the principle that sequential circuits may be tested by treating each state as a separate combinational circuit. In order to obtain the fullest possible test coverage, these procedures must be followed

- test type-1 faults by enabling one AND-term at a time
- test type-2 faults by negating one input at a time for each AND-term
- define the expected level of every output for each input vector
- use register pre-load wherever possible

Chapter 7

Design Support for PLDs

7.1 THE DESIGN PROCESS

7.1.1 CAE systems

7.1.1.1 The role of CAE

So far, we have discussed the basic ways in which the logic content of a PLD can be described, and shown how they lead to manual methods of writing this data in a format which is compatible with PLD structures. Very often though, a designer is practised in the art (or science?) of designing logic systems with standard circuits, and he should not be compelled to change his design methods in order to use a different device. If he is designing a system it is likely that he will specify its function first, with little regard to the type of device he will use to build it. Apart, that is, from the major components such as microprocessors and other dedicated LSI circuits which define the overall architecture of the system.

This is sometimes called the *top-down* approach, as it starts from a view of the overall system function, then breaks it down into smaller blocks to achieve the detail necessary to define the individual devices and their interconnections. The alternative approach is *bottom-up*, which is a more piecemeal method whereby the individual blocks are designed without regard to the overall scheme, then fitted together at the end. This may lead to some elegant solutions for the individual pieces, but will often cause problems in joining the pieces together.

Much of the design process is now automated by what is known as *Computer Aided Design*, CAD, or more broadly *Computer Aided Engineering*, CAE. The latter encompasses the whole process of converting the idea for a system into a physical entity which can be put into production. This means that not only will it perform the necessary steps to ensure that the electrical performance of the system is correct, but it will also lay out the printed boards, design their interconnections and even produce drawings of the hardware into which the system will fit. One of the great advantages of using CAE is that it generates the reference paperwork for building the system. By fitting PLD design into the framework of general CAE systems it becomes an integral tool of the designer rather than a special device standing outside the standard design methods.

7.1.1.2 Description of CAE Systems

If we restrict ourselves to the logic design function, we find that CAE systems range from software for personal computers to stand-alone machines as powerful as a mini-computer. Generally speaking, the differences lie in the quality of the graphics, the speed at which data is processed, the amount of data which can be handled and the range of functions offered. The functions which are normally considered necessary for a CAE design system are:

- design entry
- logic simulation
- timing verification
- testability analysis
- fault simulation

Some CAE systems include component level simulation which is needed for analog circuits, but is not usually relevant to logic design.

We will discuss these functions in more detail, but firstly point out that the window into what the system is doing is the graphics screen. This is used to display the circuit layout, truth tables, waveform diagrams, test vectors, etc. and so needs high definition and quick response. Since the display can usually show only part of what is happening 'inside' the system, it is made capable of being moved around as if it were a physical window open on to part of a larger picture. It can also 'zoom' in to see fine detail, or out for an overall view of the whole design. Speed of response is therefore critical to a lifelike interface.

7.1.1.3 CAE functions

We can give a more detailed look to those CAE functions which we have not already described in the course of earlier chapters.

Design entry is usually carried out by *schematic capture*. The CAE system will have a library of standard logic functions; these can be called on to the display screen and interconnected by means of keyboard commands or some physical interface, such as a tablet, mouse or light pen. The circuit schematic is built up in this way until the designer believes that he has created the correct logic function. Most often the design will require more than one screen's worth of logic symbols, so the 'window' can be moved to a 'clean' area for more components. Although the hidden parts have disappeared from immediate view, they still exist in the machine's database and can be recalled at will.

The first step after data entry is usually logic simulation. The designer will know how his design should react to applied stimuli and logic simulation enables him to test his circuit. Before CAE was established this step would be undertaken by *breadboarding*. The circuit would be built on a temporary base, called a breadboard, and real signals applied in order to check that it performed correctly. Any errors could be corrected by rewiring the circuit. Logic simulation carries out this function by simulating the effect of theoretical signals on the devices stored in its design database. The result of simulation is displayed to appear the same as if a logic analyser were looking at a real breadboard, but it has the advantage that any events which are missed can be recalled from the CAE systems memory. Errors are corrected by returning to the data entry routine and modifying the circuit. This clearly saves time in making physical changes, and it removes the need for physically obtaining the devices needed for the real breadboard.

An essential ingredient in simulating logic systems is timing verification. Each device in the CAE library has its function stored in order to make simulation possible; it will also have the timing details, delay time, set-up and hold times, etc. stored. From the timing of the applied input signals, the timing verifier will calculate the time at which each subsequent signal starts, that is, the time of each output level change. The result will be displayed as a multi-channel oscilloscope picture, or as a list of event timings. It should also note timing violations and catch spikes or glitches. Because it is based on worst-case specifications it should reveal potential problems that real devices on a breadboard might hide, because they are not worst-case.

We have already discussed testability and fault simulation. Most CAE systems include the capability to assess the testability of a circuit, usually in terms of interconnection nodes which are not observable. They are also able to measure the fault coverage of a particular set of test vectors. The set of stimuli which are used to simulate the design usually form the basis of the test vectors. A first pass at fault coverage using these inputs should cover most of the type-1 faults; most of the work on test development will probably be aimed at increasing type-2 fault coverage.

7.1.2 Dedicated software

7.1.2.1 Manufacturer specific programs

In order to assist engineers to design-in their products, most PLD manufacturers have developed software which will produce fuse maps from a design input. Many of them possess some of the features of CAE systems, though usually not as powerful in terms of the number of components which can be handled. Some will also interface directly to CAE systems. This is usually achieved by the system producing a design database in the format required by the software. In these cases the software acts as a compiler for the CAE output.

In cases where the design data is entered directly into the program the data is usually in logic equation format, although some will accept truth tables or state diagrams. The following information will be required before the fuse map can be generated:

- pin list with signal names
- target device
- logic equations, truth table or state diagram
- test vectors or simulation stimuli

Each manufacturer's software has its own syntax, so design files developed for one type of device may not be transferred easily to another manufacturer. However one designs a PLD, at some stage there will be the need to type-in the design data; some of the programs have developed ways of reducing the number of characters which need to be typed. These 'shorthand' methods are usually applicable only to fairly regular circuits. For example, in 'PALASM' (a trademark of MMI), the abbreviation INPUT[0..1,0..3] represents eight inputs with signal names INPUT00, INPUT01, INPUT02, INPUT03, INPUT10, INPUT11, INPUT12, INPUT13; whilst all the logic equations of a 4-to-1 multiplexer can be written as:

OUTPUT[*m* = 0..3] = OR[*n* = 0..3] (INPUT[*m*,*n*] * BIN[*n*](SELECT[0] SELECT[1]))

This is a high-level-language way of writing sixteen AND terms, four for each of the OUTPUTs 0–3, defining the appropriate input and binary select code for each AND term.

An alternative approach is the ability to specify intermediate functions, a feature found in 'AMAZE' (a trademark of Signetics Corp). The 4-to-1 multiplexer can be simplified by specifying the following:

$SEL0 = \overline{SELECT1} * \overline{SELECT0}$
$SEL1 = \overline{SELECT1} * SELECT0$
$SEL2 = SELECT1 * \overline{SELECT0}$
$SEL3 = SELECT1 * SELECT0$

Then the equations can be written:

OUTPUT0 = INPUT00 * SEL0
+ INPUT01 * SEL1
+ INPUT02 * SEL2
+ INPUT03 * SEL3
OUTPUT1 = INPUT10 * SEL0
etc.

This does not save as much typing as the shorthand entry but does have other advantages. The intermediate functions can be used as signal names to build up complex functions which the logic compiler will expand into an AND–OR format. This can simplify the entry of circuits which are designed with intermediate functions and are being converted to a PLD. An extension of this principle is the use of intermediate functions to define states in state machines. The equations become much more intelligible if the states are referred to by meaningful names, or even numbers, rather than just an AND term. For example, the combination lock states can be defined as:

$RST = \overline{Q3} * \overline{Q2} * \overline{Q1} * \overline{Q0}$
$IN8 = \overline{Q3} * \overline{Q2} * \overline{Q1} * Q0$
$IN0 = \overline{Q3} * \overline{Q2} * Q1 * \overline{Q0}$
$IN9 = \overline{Q3} * \overline{Q2} * Q1 * Q0$
$INFR = \overline{Q3} * Q2 * \overline{Q1} * \overline{Q0}$
$INF8 = Q3 * Q2 * \overline{Q1} * Q0$
$INF0 = \overline{Q3} * Q2 * Q1 * \overline{Q0}$
$INF9 = \overline{Q3} * Q2 * Q1 * Q0$
$FAIL = Q3 * Q2$

The input combinations are written as:

$$\begin{aligned}
\mathrm{ENT8} &= \mathrm{I3} * \overline{\mathrm{I2}} * \overline{\mathrm{I1}} * \overline{\mathrm{I0}} \\
\mathrm{ENT0} &= \overline{\mathrm{I3}} * \overline{\mathrm{I2}} * \overline{\mathrm{I1}} * \overline{\mathrm{I0}} \\
\mathrm{ENT9} &= \mathrm{I3} * \overline{\mathrm{I2}} * \overline{\mathrm{I1}} * \mathrm{I0}
\end{aligned}$$

These can be combined into the following equations:

$$\begin{aligned}
&\mathrm{IN8} && := \mathrm{RST} * \mathrm{ENT8} \\
&\mathrm{IN0} && := \mathrm{IN8} * \mathrm{ENT0} \\
&\mathrm{IN9} && := \mathrm{IN0} * \mathrm{ENT9} \\
&\mathrm{INFR} && := \mathrm{RST} * \overline{\mathrm{ENT8}} \\
&\mathrm{INF8} && := \mathrm{INFR} * \mathrm{ENT8} \\
& && \quad + \mathrm{IN8} * \overline{\mathrm{ENT0}} \\
&\mathrm{INF0} && := \mathrm{INF8} * \mathrm{ENT0} \\
& && \quad + \mathrm{IN0} * \overline{\mathrm{ENT9}} \\
&\mathrm{INF9} && := \mathrm{INF0} * \mathrm{ENT9} \\
&\mathrm{FAIL} && := \mathrm{INFR} * \overline{\mathrm{ENT8}} \\
& && \quad + \mathrm{INF8} * \overline{\mathrm{ENT0}} \\
& && \quad + \mathrm{INF0} * \overline{\mathrm{ENT9}} \\
&\mathrm{UNLOCK} && = \mathrm{IN9} \\
& && \quad + \mathrm{INF9}
\end{aligned}$$

The program must assemble the macro names, as the intermediate states may be called, into output states as well as input states. It should also perform minimisation of AND terms and, in the case of flip-flops with programmable structure, arrange the equations to fit the chosen type of flip-flop. These features are found on most manufacturers' software and make the design data easier to enter and more understandable after entry.

7.1.2.2 Independent software

The main problem with relying on one manufacturer's software is that the designer is tied to just one type of PLD. To overcome this problem some independent programs have been developed which cover a larger range of devices. Design data has to be entered in the same way as in device-dedicated software, except that there is no need to specify a target device. The programs will perform the usual logic manipulation to convert the data into AND–OR format and will then ask the designer to specify the target. An attempt is then made to fit the data into the device chosen and success or failure reported. The designer can therefore try various devices from different manufacturers and select the most economical solution.

This also gives some production security. If the chosen device becomes unavailable, or uneconomic, it is a very simple process to change to an alternative merely by changing the target device in the compilation process.

Most, if not all, of the software can be run on IBM-compatible personal computers, so no change in hardware is needed to run more than one piece of software. The main difficulty is that there are syntax differences which could be confusing in changing design programs. The information is usually stored and transferred in JEDEC format so test vector information can be included for each design. The programs will normally include fault-grading based on stuck-at-1 and stuck-at-0 faults, although the speed and number of faults which can be handled is probably less than the capability of dedicated CAE systems.

7.2 PROGRAMMING EQUIPMENT

7.2.1 Options for programming

7.2.1.1 Build or buy?

The programmer is the most vital part of the design support equipment; clearly, unless one is able to program a PLD no amount of sophisticated CAE and design expertise is going to produce a device to use at the end. Most manufacturers publish a programming specification which informs the user which voltages to apply in what sequence and for how long in order to blow each fuse, or load each cell. It is probably not beyond the scope of most competent engineers to build a circuit to implement these instructions. There must also be an interface to accept the data from the design source and translate this into fuse locations. There will probably be the need to accommodate several manufacturers and package styles and, maybe, different technologies.

The programmer is not, then, such a simple piece of equipment as might appear at first sight. We must also consider the case of programming rejects; most manufacturers accept that their devices are not 100 per cent perfect for, as we have stated before, there is no way to test programmability on non-erasable PLDs. They normally offer to replace programming rejects but need to protect themselves against losses caused by equipment which does not meet their programming specification. Device manufacturers will approve equipment made by specialist companies after extensive qualification tests but, understandably, do not have the resources to test every programmer built by potential PLD users. In the long run then, it is more cost-effective and technically better to use a commercial programmer.

In addition to the simple commercial reasons above, dedicated programmer makers are usually in close touch with the device manufacturers and are able to offer programming support to new devices as soon as, or even before, the devices themselves are available.

7.2.1.2 How complex?

There is a wide choice of commercial programmers, both in terms of the number of manufacturers and the options available. At the lowest end of the range are programmers which will cater for a small range of devices and offer only a limited number of functions. We have described various families of PLD and some equipment is able to program just one family.

In particular, EPROMs are often supported by dedicated programmers; most manufacturers use a common specification so a simple equipment will cover the whole range. The simple programming algorithm for EPROMs demands a 50 ms cycle for each address location so even a '16 k' device will take over 100 s (2048 × 50 ms) to be programmed. The commonest equipment for programming EPROMs will accept eight or more devices in parallel, the so-called *gang programmer*. This is usually capable only of copying a 'master' device, so we are left with the problem of creating the master. Most microprocessor development equipment incorporates an EPROM programmer for putting the program developed into an EPROM, which then becomes a master. If the EPROM is being used in a different application then a more complex programmer will be needed.

In any design situation, apart from a microprocessor program, the programmer must be capable of receiving data, storing it and then using it as the basis for generating the fuse pattern for whatever device is being programmed. The next stage from a pure copier is a programmer with a memory and serial port which allows it to be driven from a terminal or personal computer. This is the minimum configuration which can be considered for designing PLDs. The personal computer is used for the design itself, using a software package as described above, and the fusing information then loaded into the programmer which produces the finished device. These programmers usually cater for one or just a few PLD ranges, EPROMs or PALs and PLAs, for example.

More comprehensive, but still tied to a personal computer (PC), are programmers which are driven directly from the PC data bus. They are little more than a set of programmable power supplies and analog switches. The manufacturer's specifications are stored in the computer along with the fusing information. These offer the most versatile form of programmer since they may be updated simply by changing a floppy disk; it is not the most economic solution at present as they need a PC to perform the function which is carried out by a microprocessor in a dedicated instrument. As the cost of PCs comes down they become more attractive.

Universal programmers are probably the best solution for general PLD design. They are intended to cover all programmable devices and are usually constructed in two parts. The basic chassis contains the programmable power supplies for generating the programming waveforms, together with a keyboard, display screen and memory for storing the fuse pattern. The different ranges of PLD are covered by a series of modules which plug into the chassis. These modules contain the programming specifications, usually in an EPROM, and the circuitry for switching the appropriate waveforms to the socket pins according to the fuse pattern required.

The design data can be entered from the keyboard, or via a serial port from a separate computer. A few sophisticated programmers also have a built-in compiler which can accept logic equations directly, and therefore act as a stand-alone PLD development system. They may have a built-in disk drive to enable designs to be stored for future recall. Universal programmers without the built-in compiler will usually accept data from the keyboard in truth table or fuse chart format; they therefore need CAE back-up for designers who are not comfortable with those methods of entry.

	CAE + PROGRAMMER	PC + PROGRAMMER	UNIVERSAL PROGRAMMER
DATA ENTRY	SCHEMATIC CAPTURE LOGIC EQUATIONS	LOGIC EQUATIONS TRUTH TABLE	TRUTH TABLE FUSE CHART
LOGIC MANIPULATION	BOOLEAN EXPANSION MINIMISATION	BOOLEAN EXPANSION	
DESIGN VERIFICATION	FULL SIMULATION	LIMITED SIMULATION	
TIMING VERIFICATION	HAZARD ANALYSIS WAVEFORM SYNTHESIS		
TEST GENERATION	TESTABILITY ANALYSIS AUTOTEST GENERATION	FAULT GRADING	VECTOR ENTRY

Fig. 7.1 CAE based design flows.

7.2.2 The PLD development system

We can complete this section by summarising the options for creating a PLD design and the hardware needed to accomplish this. Figure 7.1 shows the design flow and the capabilities of the three possible hardware configurations: a full CAE system plus programer, a PC running dedicated software plus programmer, and a universal programmer. The CAE system could be a PC-oriented system, but with universal capability in terms of design entry and simulation; dedicated software implies a capability for handling PLD design only.

It is apparent that as one moves from a universal system towards a dedicated PLD system the range of facilities diminishes. Thus, while CAE offers schematic capture or logic equation entry and full simulation and test vector analysis, a universal programmer is limited to truth table or fuse chart entry, and the ability to enter and store test vectors. Probably the best compromise is to install a universal design program and use this to drive a universal programmer. It would then be possible to add CAE tools, such as schematic capture, when the need arose. This presupposes that a full CAE system is not already installed.

7.2.3 Production programming

Having successfully completed a design the need arises for production programming. There are two ways in which this can be undertaken: the production/test department of the company can undertake the work, or the supplier of the PLDs can program at source. As with most of the situations encountered so far, there are arguments for and against both approaches.

Keeping the programming in-house means that a better control of stock and schedules can be maintained. If the same type of PLD is used in more than one application then a single purchase can be made and the production planning made more in line with last-minute requirements. It should also lead to a wider choice of PLD supplier since one is not restricted to buying from the company that has details of your design. Programming in-house is also likely to be cheaper, provided that quantities are fairly high, because you are paying only for your own labour instead of a profit margin on somebody else's labour.

Against that, there is the problem of rejects which, although they are usually replaced, mean ordering extra product and raising extra paperwork to return them. In most cases the production department will need its own programmer so that the design and production can be independent of each other. This will attract an overhead of something in the order of £1000 p.a. ($1400 p.a.) causing an added cost of 10 p (14 c) plus labour at a usage rate of 10000 devices per year. The breakeven point compared with an outside programming facility depends on the relative labour charge compared with equipment overhead, which will be much less for the outsider programming much larger quantities.

The argument thus revolves around economics again and will clearly depend on individual circumstances. There is certainly a case for considering buying ready programmed, or subcontracting the programming, unless the volume involved is in the thousands. In that case it may be worth considering moving on from PLDs as we shall discuss next.

7.3 DESIGNING-OUT PLDs

7.3.1 Problems of scale

So far, we have tended to look at PLDs as the only solution to implementing logic, although we have mentioned the other approaches: standard circuits, gate arrays, cell arrays, and full custom. In Chapter 5 we looked at the economics of replacing standard circuits with PLDs and found that a replacement factor of from 3 to 6, or higher for more complex PLDs, was likely to be economic. A similar sum can be calculated for comparing PLDs with other ASICs (*Application-Specific ICs*). If we do this we have to add another factor to the cost of ownership, that is the start-up or design cost of a masked ASIC.

Even assuming that the customer does most of the work, the mask costs for a gate array are likely to be at least £5000 ($7000), and the designer would need a full CAE system to implement the design. One gate array might replace five PLDs, although this is dependent entirely on the type of PLD and size of gate array but, with this assumption, a thousand arrays at £5 ($7) would cost the same as the equivalent circuit in PLDs at £2 ($2.8). There are other factors to be taken into account: time to samples may be three to four weeks for a gate array, mistakes cost another mask charge, and a commitment to full production must be made. On the other hand, assembly costs will be lower and performance probably better for the array. Overall, there will be a breakpoint of at least a thousand where a masked ASIC becomes the preferred solution.

Although this book is an exhortation to use PLDs as the ideal way of building logic circuits, considering ease of design and economics as the prime motivations, this section will consider PLDs as stepping-stones to even more cost-effective solutions. The potential user must perform a cost analysis based on the above arguments to decide which is the optimum route for him.

7.3.2 Hard Array Logic

Hard Array Logic, or HAL (a trademark of MMI), was introduced as an answer to the problem outlined above: that masked devices will generally be cheaper than programmable, once a threshold volume has been passed. PLDs, as we have seen, need additional circuitry to allow the fuse array to be addressed from the input pins, effectively bypassing the logic circuit. This is the main reason why they are more expensive than their masked equivalents. The other reason for them being less preferable is the fact that they need programming. This is a distinct advantage in the start-up phase of a design: it allows changes to be introduced painlessly, and reduces the risks involved in holding component stocks. Once a product is established in terms of design and run rate, programming can become an expensive and irksome additional process.

The HAL has the same basic stucture as its equivalent PAL, but the logic function is defined by a mask instead of by fuses. It is otherwise a complete replacement part in terms of pinning, function and performance. A PAL can be used for the design and early production phase of a product but once the product is established all, or part, of the production can be taken over by a HAL. Like all custom-masked devices, supply quantities of HALs have to be committed several months ahead; if the HAL supply is planned conservatively then any extra product can be met by PALs. This gives the user the best of both worlds: the cost savings of a masked ASIC with the volume flexibility of a programmable device.

7.3.3 Conversion to a gate array

If a product is being designed with the intention, or at least expectation, that it will be produced in high volume, then it is likely that the random logic will be designed into a masked ASIC. The initial steps of the design are the same as for a PLD, that is, logic input and design verification. Thus, there is no more work involved in designing the circuit as a PLD than there would be in going straight to a gate array or cell array. There is the advantage that a device can be available within minutes of completing the design, compared with the weeks usually needed for masked samples to appear.

Instant availability of hardware to test is likely to be a benefit to the designer but is there a price to pay? If the design data has to be entered twice, then clearly there is, but if the same database can be used for both, then there is no problem. Many of the CAE system manufacturers now offer the facility of using standard PLD software files as a design input. The CAE will convert these to gate- or cell-based primitives for the target array just as it does for standard logic devices.

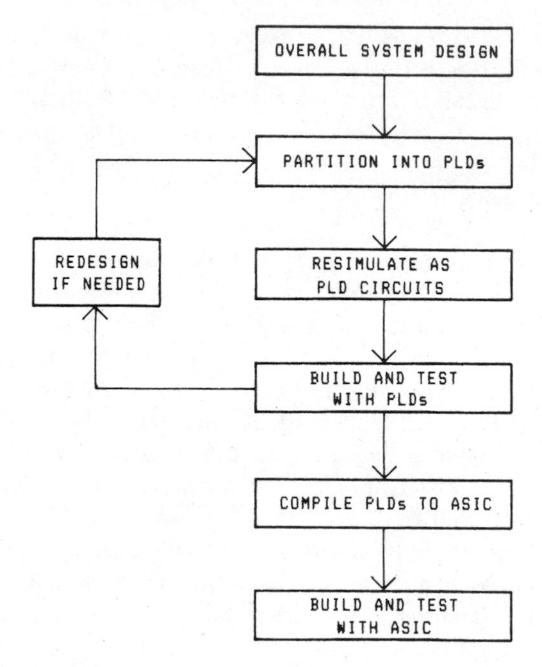

Fig. 7.2 ASIC design flow using PLDs as an intermediate stage.

PLDs can therefore be used as a powerful design tool for masked-array designers. They also provide hardware back-up in case there is a failure with the masked array.

The design flow, using PLDs as an intermediate stage, then appears as in Figure 7.2. In a top-down design the functions intended for the masked device must be partitioned for PLDs. As these are not intended to go into production, a 'minimum cost' solution is not necessary; a 'minimum wiring' solution may even be preferred. The design data for each PLD can then be entered; when all the PLDs have been entered the data may be transferred to the array design and the logic simulated as a total entity. Any necessary changes can be made to the individual blocks and resimulated until the overall function is correct. The design file can then be used for laying out the array, while the PLD files are used for creating devices containing the required logic.

The system can be built and tested before the array samples are produced, or even before any financial commitment is made to producing masks. It also becomes possible to build prototypes for demonstration or early production, and effectively come to the market quicker by using this approach. Some of the risks involved in using masked ASICs are substantially reduced if PLDs are used as an intermediate stage in the design; the additional cost and work is insignificant in comparison to the benefits obtained.

Chapter 8

Programmable Logic Applications

8.1 COMBINATIONAL MACRO ELEMENTS

8.1.1 Introduction

8.1.1.1 The purpose of macros

Applications information can be presented in two ways, each with its own use. One way is to look at the various uses to which devices have been put, the purpose being to trigger ideas in the mind of the designer who may be contemplating similar circuits. The second way is to provide the designer with building bricks, called *macros*, which enable him to synthesise circuits as he might do from standard logic families. This is the approach in this section, the first part of which covers the combinational circuits. Because most designers are familiar with 74-series and 4000-series numbering, these are used as the bases of the macro descriptions.

Four methods of describing the macros are used, where appropriate: logic equations, Karnaugh map, truth table, and PAL fuse chart. The macro descriptions can therefore be used by a designer using any of the standard methods of entering logic information into a PLD.

8.1.2 Simple gates

8.1.2.1 AND gates

74-series types: 08, 11, 21
4000-series types: 4073, 4081, 4082

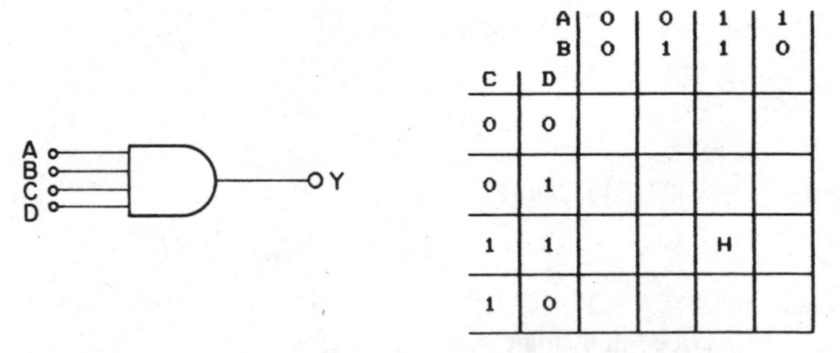

Fig. 8.1 AND symbol and Karnaugh map.

Logic equation: Y = A * B * C * ...

Truth table:

Active level –				*H*
A	*B*	*C*	...	*Y*
H	H	H	...	A

X-chart: X– X– X– ... (use PAL*mHn*)

8.1.2.2 NAND gates

74-series types: 00, 10, 20, 30, 133
4000-series types: 4011, 4012, 4023, 4068

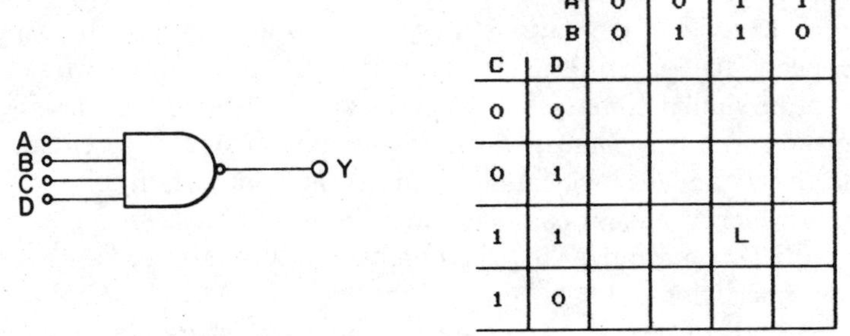

Fig. 8.2 NAND symbol and Karnaugh map.

Logic equation: $\overline{\mathrm{Y}}$ = A * B * C * ...

Truth table:

Active level –				*L*
A	*B*	*C*	...	*Y*
H	H	H	...	A

X-chart: X– X– X– ... (use PAL*mLn*)

8.1.2.3 OR gates

74-series type: 32
4000-series types: 4071, 4072, 4075

Logic Equation: Y = A + B + C + ...
This may be described in a single AND term as:

$$\overline{\mathrm{Y}} = \overline{\mathrm{A}} * \overline{\mathrm{B}} * \overline{\mathrm{C}} * \ldots$$

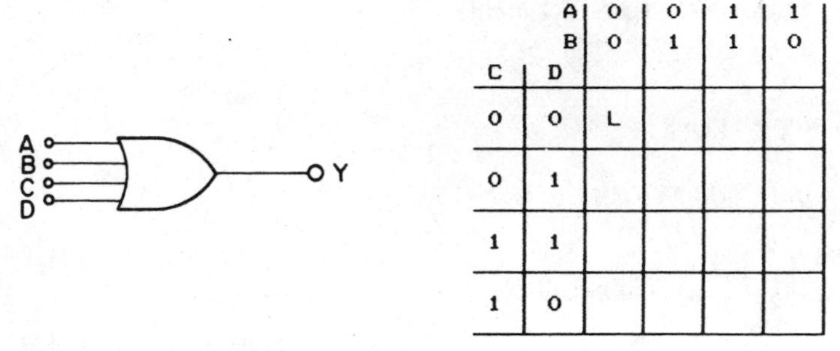

Fig. 8.3 OR symbol and Karnaugh map.

Truth Table:

Active level –				*L*
A	*B*	*C*	...	*Y*
L	L	L	...	A

X-chart: –X –X –X ... (use PAL*mLn*)

8.1.2.4 NOR gates

74-series types: 02, 27
4000-series types: 4000, 4001, 4002, 4025, 4078

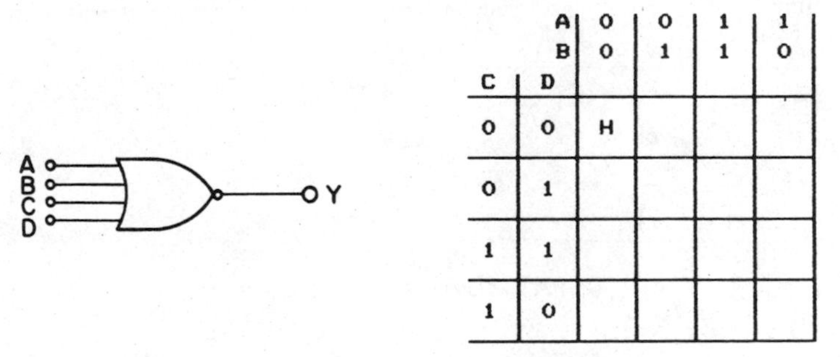

Fig. 8.4 NOR symbol and Karnaugh map.

Logic equation: $\overline{Y} = A + B + C + \ldots$
This may be described in a single AND term as:

$$Y = \overline{A} * \overline{B} * \overline{C} * \ldots$$

Truth table:

Active level –				*H*
A	*B*	*C*	...	*Y*
L	L	L	...	A

X-chart: –X –X –X ... (use PAL*mHn*)

8.1.3 Complex gates

8.1.3.1 AND–OR–INVERT gates

74-series types: 51, 54, 64
4000-series types: 4019, 4085, 4086, 4506

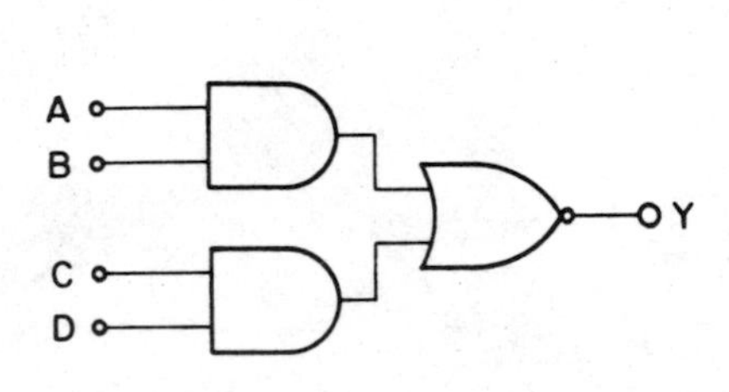

C	D	A=0 B=0	A=0 B=1	A=1 B=1	A=1 B=0
0	0			L	
0	1			L	
1	1	L	L	L	L
1	0			L	

Fig. 8.5 AOI symbol and Karnaugh map.

Logic equation: $\overline{Y} = A * B + C * D$

Truth table:

Active level –				*L*
A	*B*	*C*	*D*	*Y*
H	H	–	–	A
–	–	H	H	A

X-chart: X– X– -- -- (use PAL*mLn*)
-- -- X– X–

8.1.3.2 Exclusive-OR gates

74-series types: 86, 135
4000-series types: 4030, 4070, 4077, 4507

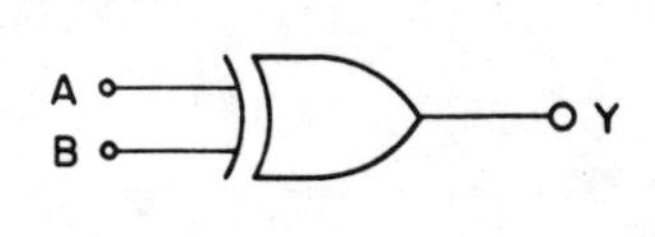

B	A=0	A=1
0	L	H
1	H	L

Fig. 8.6 Exclusive-OR symbol and Karnaugh map.

Logic Equation: Y = A :+: B

This may be expanded as:

Either $Y = A * \overline{B} + \overline{A} * B$

Or $\overline{Y} = A * B + \overline{A} * \overline{B}$

The respective truth tables are:

Active level –		*H*
A	*B*	*Y*
H	L	A
L	H	A

Active level –		*L*
A	*B*	*Y*
H	H	A
L	L	A

The corresponding X-charts are:

X– –X (use PAL*mHn*)
–X X–

X– X– (use PAL*mLn*)
–X –X

8.1.4 Controlled outputs

8.1.4.1 Tri-state buffer or gate

74-series types: 125, 126, 134, 240, 241, 244
4000-series types: 40097, 40098, 4502, 4503

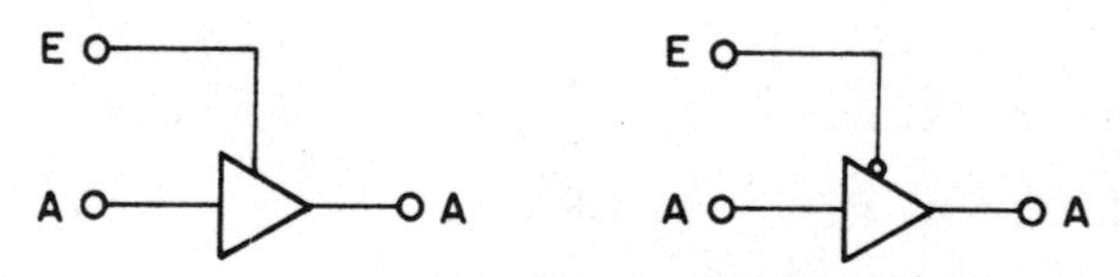

Fig. 8.7 Tri-state buffer symbols.

Tri-state operation is specified in logic equations by means of a condition applied to the output, thus:

if(E) Y = A * ....

For an active-LOW enable this becomes:

if($\overline{E}$) Y = A * ....

To unconditionally enable a tri-state the convention is:

if(V_{cc}) Y = A * ....

The tri-state control terms in a truth table are usually collected in a separate section below the logic terms. In the following table, outputs 3 and 2 have active-HIGH control, outputs 1 and 0 have active-LOW control:

	E	*A*	*B*	....	*Y3 Y2 Y1 Y0*
D3	H	–	–	....	→ Y3
D2	H	–	–	....	→ Y2
D1	L	–	–	....	→ Y1
D0	L	–	–	....	→ Y0

In an X-chart, the tri-state control term is usually the first in any group of AND terms, as with the active-HIGH and active-LOW shown respectively below:

X– –– –– ....
–– X– ....
.
.
–X –– –– ....
–– X– ....

8.1.4.2 Open collector outputs

74-series types: 01, 03, 26, 33

Open collector outputs, as such, are available on only a few PLEs and very few PLAs. However, a quasi open collector output can be formed with a tri-state output. The condition to be defined is that the output is LOW when the logic is true, and high-impedance when the logic is not true. Because the tri-state must be defined with a single AND term only the NAND and OR functions can be given an 'open collector' in this way. The equations for open collector NAND and open collector OR are resectively:

if(A * B * C * ....) $\overline{\text{Y}}$ = A * B * C * ....
if($\overline{\text{A}}$ * $\overline{\text{B}}$ * $\overline{\text{C}}$ * ....) $\overline{\text{Y}}$ = $\overline{\text{A}}$ * $\overline{\text{B}}$ * $\overline{\text{C}}$ * ....

The truth table for NAND (Y1) and OR (Y0) is:

Active Level L					*L*
	A	*B*	*C....*	*Y1*	*Y0*
	H	H	H....	A	.
	L	L	L....	.	A
	.				
	.				
D1	H	H	H....	→ Y1	
D0	L	L	L....		→ Y0

X-chart for NAND (upper group) and OR (lower group) is:

X– X– X– ....
X– X– X– ....
XX XX XX ....
.
.
–X –X –X ....
–X –X –X ....
XX XX XX ....

8.1.4.3 Transceivers

74-series types: 242, 243, 245

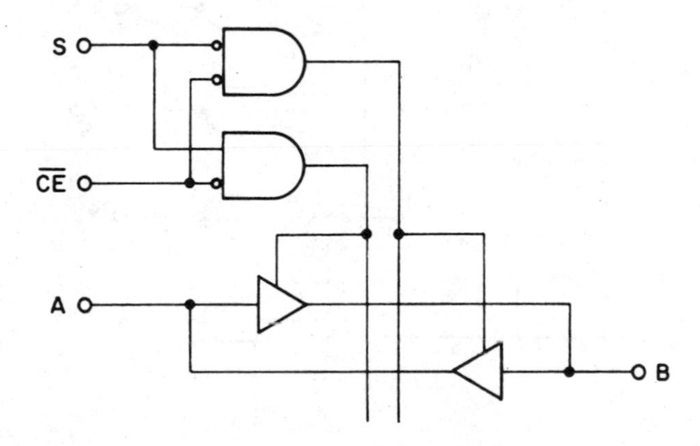

Fig. 8.8 Transceiver symbol.

Logic Equations: if($\overline{\text{CE}} * \text{S}$) B = A
if($\overline{\text{CE}} * \overline{\text{S}}$) A = B

To build this function, A and B must be bidirectional pins; AI and BI refer to A and B in the AND array, while AO and BO refer to the OR array in the following truth table:

	Active Level				*H*	*H*
	CES	*AI*	*BI*	*AO*	*BO*	
	–	–	H	–	.	A
	–	–	–	H	A	.
	.					
	.				↑	↑
D1	L	H	–	–	⌟	
D0	L	L	–	–	————	⌟

In the following X-chart, output A is the upper group and output B the lower:

CE	*S*	*A*	*B*
–X	X–	--	--
--	--	--	–X
XX	XX	XX	XX
.			
.			
–X	–X	--	--
--	--	–X	--
XX	XX	XX	XX

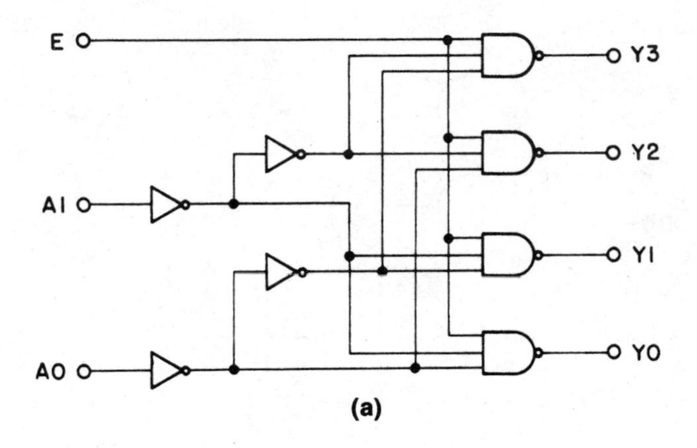

(a)

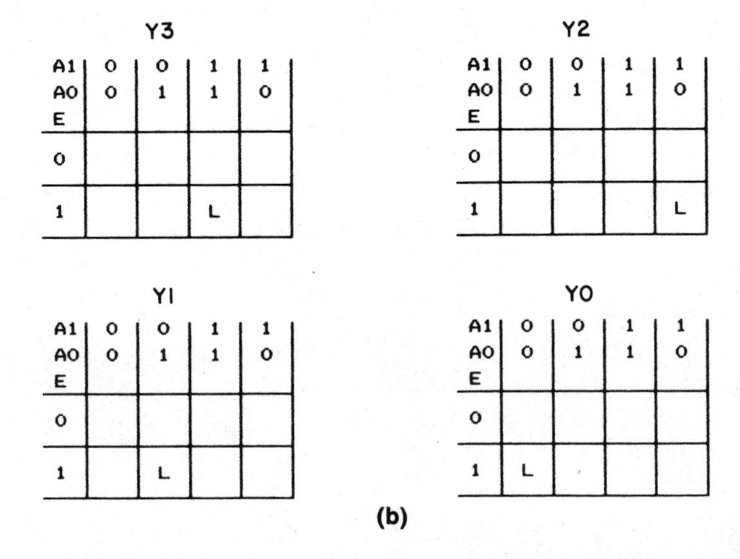

(b)

Fig. 8.9 1-of-2 decoder circuit and Karnaugh map.

An active-LOW PAL is assumed as these are more common among 'feedback' types.

8.1.5 Decoders, encoders and multiplexers

8.1.5.1 Decoders

74-series types: 42, 138, 139, 154, 155, 156
4000-series types: 4514, 4515, 4528, 4555, 4556

Logic equations:

$\overline{Y0} = E * \overline{A1} * \overline{A0}$
$\overline{Y1} = E * \overline{A1} * A0$
$\overline{Y2} = E * A1 * \overline{A0}$
$\overline{Y3} = E * A1 * A0$

Truth table:

E	*A1*	*A0*	*Y3*	*Y2*	*Y1*	*Y0*
Active Level			*L*	*L*	*L*	*L*
H	L	L	.	.	.	A
H	L	H	.	.	A	.
H	H	L	.	A	.	.
H	H	H	A	.	.	.

X-chart:

E	*A1*	*A2*			
X–	X–	X–	––	....	(Y3)
.					
X–	X–	–X	––	....	(Y2)
.					
X–	–X	X–	––	....	(Y1)
.					
X–	–X	–X	––	....	(Y0)

Because each output requires only a single AND term an address decoder, such as PLS151 or PAL6L16, can be used for this function. The above design is for active-LOW outputs and would therefore require a PAL*m*L*n* type. It could be changed to active-HIGH by changing active levels to 'H' or using a PAL*m*H*n*.

8.1.5.2 Priority encoders

74-series types: 147, 148
4000-series types: 40147, 4532

I3	I2	I1	I0	A1	A0
L	X	X	X	H	H
H	L	X	X	H	L
H	H	L	X	L	H
H	H	H	L	L	L

A1

I1	I3 / I2 / I0	0 / 0	0 / 1	1 / 1	1 / 0
0	0			L	
0	1			L	
1	1				
1	0			L	

A0

I1	I3 / I2 / I0	0 / 0	0 / 1	1 / 1	1 / 0
0	0				L
0	1				L
1	1				L
1	0			L	L

Fig. 8.10 4-input priority encoder function table and Karnaugh map.

Logic equations (by reference to the Karnaugh maps we can implement a glitch-free design by overlapping the AND terms):

$$\overline{A1} = I3 * I2 * \overline{I1} + I3 * I2 * \overline{I0}$$
$$\overline{A0} = I3 * \overline{I2} + I3 * I1 * \overline{I0}$$

Truth table:

I3	*I2*	*I1*	*I0*	*A1*	*A0*
	Active Level			*L*	*L*
H	H	L	–	A	.
H	H	–	L	A	.
H	L	–	–	.	A
H	–	H	L	.	A

X-chart (PAL*mLn*):

I3	*I2*	*I1*	*I0*		
X–	X–	–X	––	....	($\overline{A1}$)
X–	X–	––	–X	....	
.					
.					
X–	–X	––	––	....	($\overline{A2}$)
X–	––	X–	–X	....	

8.1.5.3 Multiplexers

74-series types: 150, 151, 153, 157, 158
4000-series types: 4019, 40257, 4539

S1	S0	I3	I2	I1	I0	Y
L	L	X	X	X	H	H
L	L	X	X	X	L	L
L	H	X	X	H	X	H
L	H	X	X	L	X	L
H	L	X	H	X	X	H
H	L	X	L	X	X	L
H	H	H	X	X	X	H
H	H	L	X	X	X	L

		S1	0	0	0	0	1	1	1	1
		S0	0	0	1	1	1	1	0	0
		I3	0	1	1	0	0	1	1	0
I2	I1	I0								
0	0	0						H		
0	0	1	H	H				H		
0	1	1	H	H	H	H		H		
0	1	0			H	H		H		
1	1	0			H	H		H	H	H
1	1	1	H	H	H	H		H	H	H
1	0	1	H	H				H	H	H
1	0	0						H	H	H

Fig. 8.11 4-input multiplexer function table and Karnaugh map.

Logic equation (referring to the Karnaugh map shows that the minimum solution of four AND terms does not produce any overlapping, so this solution is prone to hazards):

$$
\begin{aligned}
Y = {} & S1 * S0 * I3 \\
& + S1 * \overline{S0} * I2 \\
& + \overline{S1} * S0 * I1 \\
& + \overline{S1} * \overline{S0} * I0
\end{aligned}
$$

Truth table:

S1	*S0*	*I3*	*I2*	*I1*	*I0*	*Y* Active Level *H*
H	H	H	–	–	–	A
H	L	–	H	–	–	A
L	H	–	–	H	–	A
L	L	–	–	–	H	A

X-chart (PAL*mHn*):

S1	*S0*	*I3*	*I2*	*I1*	*I0*	
X–	X–	X–	--	--	--	....
X–	–X	--	X–	--	--	....
–X	X–	--	--	X–	--	....
–X	–X	--	--	--	X–	....

A2	A1	A0	B2	B1	B0	A>B	A=B	A<B
H	X	X	L	X	X	H	L	L
H	H	X	H	L	X	H	L	L
L	H	X	L	L	X	H	L	L
H	H	H	H	H	L	H	L	L
H	L	H	H	L	L	H	L	L
L	H	H	L	H	L	H	L	L
L	L	H	L	L	L	H	L	L
H	H	H	H	H	H	L	H	L
H	H	L	H	H	L	L	H	L
H	L	H	H	L	H	L	H	L
etc.								
L	X	X	H	X	X	L	L	H
H	L	X	H	H	X	L	L	H
L	L	X	L	H	X	L	L	H
etc.								

		A2	0	0	0	0	1	1	1	1
		A1	0	0	1	1	1	1	0	0
		A0	0	1	1	0	0	1	1	0
B2	B1	B0								
0	0	0	E	G	G	G	G	G	G	G
0	0	1	S	E	G	G	G	G	G	G
0	1	1	S	S	E	S	G	G	G	G
0	1	0	S	S	G	E	G	G	G	G
1	1	0	S	S	S	S	E	G	S	S
1	1	1	S	S	S	S	S	E	S	S
1	0	1	S	S	S	S	G	G	E	S
1	0	0	S	S	S	S	G	G	G	E

Fig. 8.12 Magnitude comparator function table and Karnaugh map.

8.1.6 Arithmetic circuits

8.1.6.1 Magnitude comparator

74-series type: 85
4000-series types: 4063, 40085, 4585

Logic equations:
The Karnaugh map is drawn in terms of A>B (G), A=B (E) and A<B (S), from which it may be seen that 'E' is just $\overline{G} \ast \overline{S}$, and that 'G' and 'S' are interchangeable by exchanging 'A' and 'B'. We will just write the equation for the 'G'-terms as:

$$
\begin{aligned}
A>B = {} & \overline{A2} \ast A0 \ast \overline{B2} \ast \overline{B1} \ast \overline{B0} \\
& + A1 \ast \overline{B2} \ast \overline{B1} \\
& + A2 \ast \overline{B2} \\
& + A2 \ast A1 \ast \overline{B1} \\
& + A2 \ast A0 \ast \overline{B1} \ast \overline{B0} \\
& + A2 \ast A1 \ast A0 \ast B1 \ast \overline{B0} \\
& + \overline{A2} \ast A1 \ast A0 \ast \overline{B2} \ast \overline{B0}
\end{aligned}
$$

Truth table:

								Bidirectional Pins		
						Inputs		*Outputs*		
						Active Level		*H*	*H*	*H*
A2	*A1*	*A0*	*B2*	*B1*	*B0*	*A>B*	*A<B*	*A>B*	*A<B*	*A=B*
L	–	H	L	L	L	–	–	A	.	.
–	H	–	L	L	–	–	–	A	.	.
H	–	–	L	–	–	–	–	A	.	.
H	H	–	–	L	–	–	–	A	.	.
H	–	H	–	L	L	–	–	A	.	.
H	H	H	–	H	L	–	–	A	.	.
L	H	H	L	–	L	–	–	A	.	.
–	–	–	–	–	–	L	L	.	.	A
.										
.										

X-chart (PAL*m*H*n*):

A2	*A1*	*A0*	*B2*	*B1*	*B0*	*A>*	*A<*	
-X	--	X-	-X	-X	-X	--	--	(A>B)
--	X-	--	-X	-X	--	--	--	
X-	--	--	-X	--	--	--	--	
X-	X-	--	--	-X	--	--	--	

A2	*A1*	*A0*	*B2*	*B1*	*B0*	*A>*	*A<*	
X–	––	X–	––	–X	–X	––	––	
X–	X–	X–	––	X–	–X	––	––	
–X	X–	X–	–X	––	–X	––	––	
.								
.								
––	––	––	––	––	––	–X	–X	(A=B)

8.1.6.2 Parity generator

74-series types: 180, 280
4000-series types: 40101, 4531

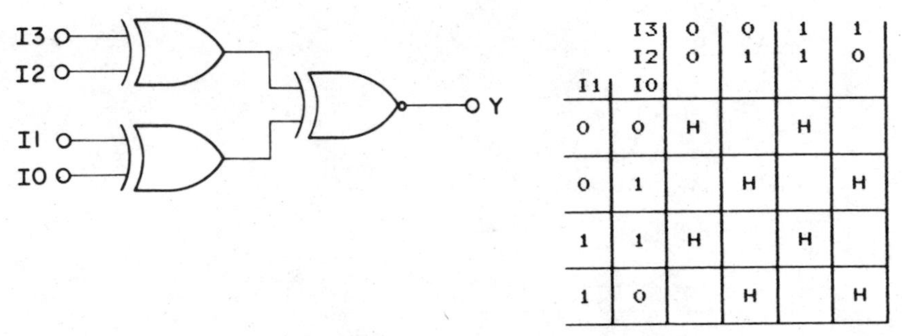

Fig. 8.13 4-input parity generator circuit and Karnaugh map.

Logic equation: from the Karnaugh map a separate AND term is required for each 'H'-cell, that is, eight altogether, and this total doubles for each additional input. A less AND term intensive design which uses more outputs is to construct the exclusive-OR gates and interconnect these. This structure uses up outputs except in those devices which possess internal feedback paths. The equation, for even parity, is:

PE = (I3:+: I2) :+: (I1 :+: I0)

The brackets are optional, and are included merely for clarity

Truth table:

				Bidirectional Pins				
				Inputs		*Outputs*		
I3	*I2*	*I1*	*I0*	*X1*	*X0*	*X1*	*X0*	*PE*
			Active Level			*H*	*H*	*H*
H	H	–	–	–	–	A	.	.
L	L	–	–	–	–	A	.	.
–	–	H	H	–	–	.	A	.
–	–	L	L	–	–	.	A	.
–	–	–	–	H	H	.	.	A
–	–	–	–	L	L	.	.	A

X-chart (PAL*m*H*n*):

I3	*I2*	*I1*	*I0*	*X1*	*X0*	
X–	X–	--	--	--	--	(X1)
–X	–X	--	--	--	--	
.						
--	--	X–	X–	--	--	(X0)
--	--	–X	–X	--	--	
.						
--	--	--	--	X–	X–	(PE)
--	--	--	--	–X	–X	

8.1.6.3 Full adder

74-series types: 83, 283
4000-series types: 4008, 4568(BCD)
Analysis of the Karnaugh maps shows that, even with AND term sharing, as in a PLA structure, sixteen AND terms are needed to implement this function, which is only half as complex as a standard logic device. In the standard logic families much use is made of exclusive-OR gates, which, as we saw with the parity generator, need to use a bidirectional pin and are thus of limited use in PLDs. It is better to use PROMs, or PLEs, to construct arithmetic functions in PLDs. It is usually possible to generate the address/data code for a PLE with a computer for arithmetic functions; as an example we will list these for the two-bit adder.

	C	0	0	0	0	1	1	1	1
	A1	0	0	1	1	1	1	0	0
	A0	0	1	1	0	0	1	1	0
B1	B0								
0	0			H	H	H		H	
0	1				H			H	H
1	1	H		H		H	H		
1	0	H	H				H		H

SI

	C	0	0	0	0	1	1	1	1
	A1	0	0	1	1	1	1	0	0
	A0	0	1	1	0	0	1	1	0
B1	B0								
0	0		H	H		H			H
0	1	H			H		H	H	
1	1	H			H		H	H	
1	0		H	H		H			H

SO

Fig. 8.14 2-bit adder Karnaugh map.

	C	0	0	0	0	1	1	1	1
	A1	0	0	1	1	1	1	0	0
	A0	0	1	1	0	0	1	1	0
B1	B0								
0	0						H		
0	1			H		H	H		
1	1		H	H	H	H	H	H	H
1	0			H	H	H	H	H	

C

Fig. 8.14 2-bit adder Karnaugh map. (*Cont.*)

The pin/function assignment is:

C_{in} – A4	C_{out} – Q2
B1 – A3	S1 – Q1
B0 – A2	S0 – Q0
A1 – A1	
A0 – A0	

The address/data table, in hexadecimal, is:

	0	1	2	3	4	5	6	7	8	9	A	B	C	D	E	F
0000	0	1	2	3	1	2	3	4	2	3	4	5	3	4	5	6
0010	1	2	3	4	2	3	4	5	3	4	5	6	4	5	6	7

8.1.7 Latches

8.1.7.1 D-latch

74-series types: 75, 118, 373
4000-series types: 4042, 40373, 4532

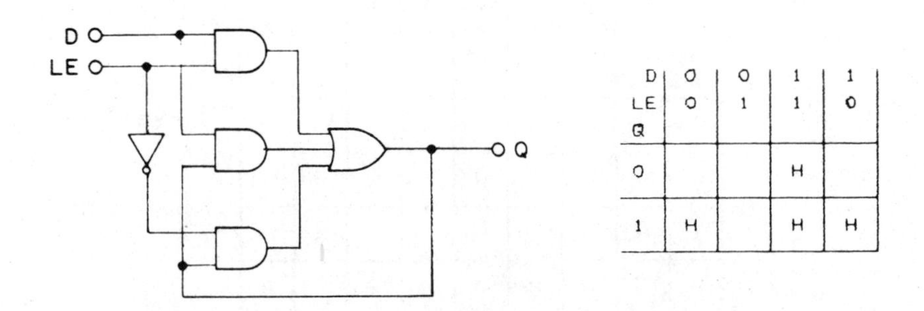

Fig. 8.15 D-latch circuit and Karnaugh map.

Logic equation:

$$Q = D * LE + Q * \overline{LE} + D * Q$$

Truth table:

		Bid.	Pins	
		I/P	*O/P*	
D	*LE*	*Q*	*Q*	
Active Level H				
H	H	–	A	
–	L	H	A	
H	–	H	A ('deglitch' term)	

X-chart (PAL*m*H*n*):

D	*LE*	*Q*
X–	X–	––
––	–X	X–
X–	––	X–

8.1.7.2 R–S Latches

74-series type: 279
4000-series types: 4043, 4044

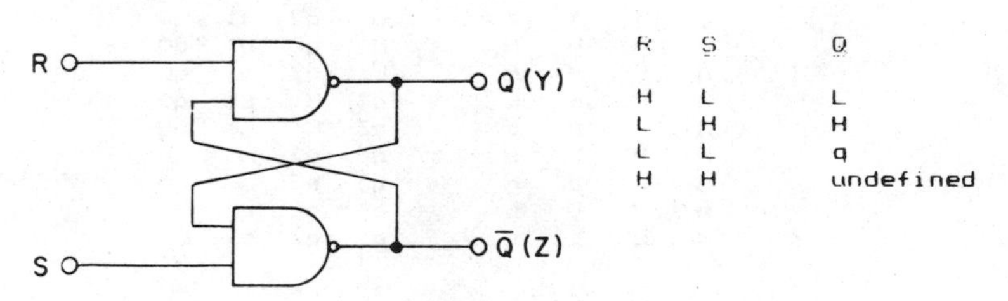

Fig. 8.16 R-S latch circuit and function table.

Logic equations:

$$Q = (R * \overline{Q})$$
$$\overline{Q} = (\overline{S * Q})$$

There may be some confusion here in that Q and $\overline{Q}$ are **different** outputs, and not merely complemented versions of one output. In most CAD systems Q and $\overline{Q}$ will have to be given different symbols, for example Y and Z, and the equations then become:

$$\overline{Y} = R * Z$$
$$\overline{Z} = S * Y$$

Truth table:

		Bidirectional Pins			
		Inputs		*Outputs*	
R	*S*	*Q*	$\overline{Q}$	*Q*	$\overline{Q}$
Active Level				*L*	*L*
H	–	–	H	A	.
–	H	H	–	.	A

X-chart (PAL*m*L*n*):

R	*S*	*Q*	$\overline{Q}$	
X–	--	--	X–	(Q)
.				
.				
--	X–	X–	--	($\overline{Q}$)

8.1.7.3 Addressable latch

74-series types: 256, 259
4000-series types: 4099, 4724

E	D	A1	A0	Q3	Q2	Q1	Q0
L	X	X	X	q3	q2	q1	q0
H	H	H	H	H	q2	q1	q0
H	L	H	H	L	q2	q1	q0
H	H	H	L	q3	H	q1	q0
H	L	H	L	q3	L	q1	q0
H	H	L	H	q3	q2	H	q0
H	L	L	H	q3	q2	L	q0
H	H	L	L	q3	q2	q1	H
H	L	L	L	q3	q2	q1	L

	E	0	0	0	0	1	1	1	1
	D	0	0	1	1	1	1	0	0
	Q3	0	1	1	0	0	1	1	0
A1	A0								
0	0		H	H			H	H	
0	1		H	H			H	H	
1	1		H	H		H	H		
1	0		H	H			H	H	

Q3

Fig. 8.17 4-bit addressable latch function table and Karnaugh map (Q3 only).

Logic equation (Q3 only):

$$
\begin{aligned}
Q3 &= A1 * A0 * E * D \\
&+ \overline{E} * Q3 \\
&+ \overline{A1} * E * Q3 \\
&+ \overline{A0} * E * Q3 \\
&+ D * Q3 \text{ (deglitch term)}
\end{aligned}
$$

Truth table:

A1	*A0*	*E*	*D*	*I/P Q3*	*O/P Q3*
	Active Level				*H*
H	H	H	H	–	A
–	–	L	–	H	A
L	–	H	–	H	A
–	L	H	–	H	A
–	–	–	H	H	A

X-chart (PAL*m*H*n*):

A1	*A0*	*E*	*D*	*Q3*	*Q2*	....
X–	X–	X–	X–	––	––	....
––	––	–X	––	X–	––	....
–X	––	X–	––	X–	––	....
––	–X	X–	––	X–	––	....
––	––	––	X–	X–	––	....

8.1.8 Flip-flops

8.1.8.1 D-type flip-flop

74-series types: 74, 173, 174, 175, 273, 374
4000-series types: 4013, 40174, 40175, 40374

Logic equations:

$$
\begin{aligned}
X &= \overline{CK} * D * \overline{RS} \\
&+ X * D * \overline{RS} \\
&+ CK * X * \overline{RS} \\
&+ PR
\end{aligned}
$$

$$
\begin{aligned}
Q &= CK * X * \overline{RS} \\
&+ Q * X * \overline{RS} \\
&+ \overline{CK} * Q * \overline{RS} \\
&+ PR
\end{aligned}
$$

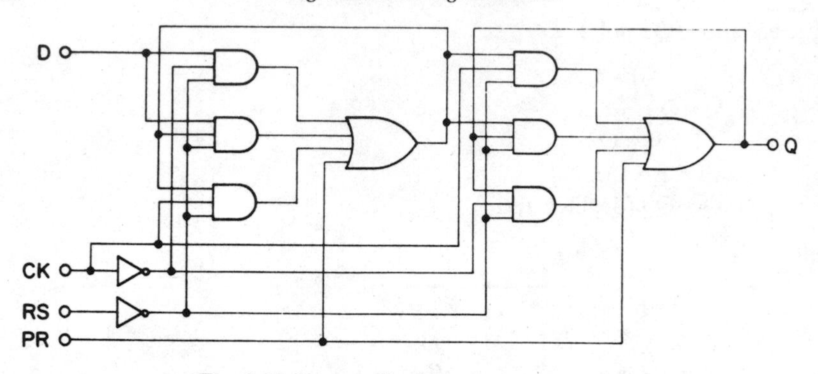

Fig. 8.18 D-type flip-flop circuit diagram.

Truth table:

				Bidirectional Pins			
				Inputs		*Outputs*	
D	*CK*	*PR*	*RS*	*X*	*Q*	*X*	*Q*
			Active Level			*H*	*H*
H	L	–	L	–	–	A	.
H	–	–	L	H	–	A	.
–	H	–	L	H	–	A	.
H	H	H	–	–	–	A	A
–	H	–	L	H	–	.	A
–	–	–	L	H	H	.	A
–	L	–	L	–	H	.	A

X-chart (PAL*m*H*n*):

D	*CK*	*PR*	*RS*	*X*	*Q*
X–	–X	––	–X	––	––
X–	––	––	–X	X–	––
––	X–	––	–X	X–	––
––	––	X–	––	––	––
.					
.					
––	X–	––	–X	X–	––
––	––	––	–X	X–	X–
––	–X	––	–X	––	X–
––	––	X–	––	––	––

8.1.8.2 J–K flip-flop

74-series types: 107, 109, 112, 113
4000-series types: 4027, 4095, 4096

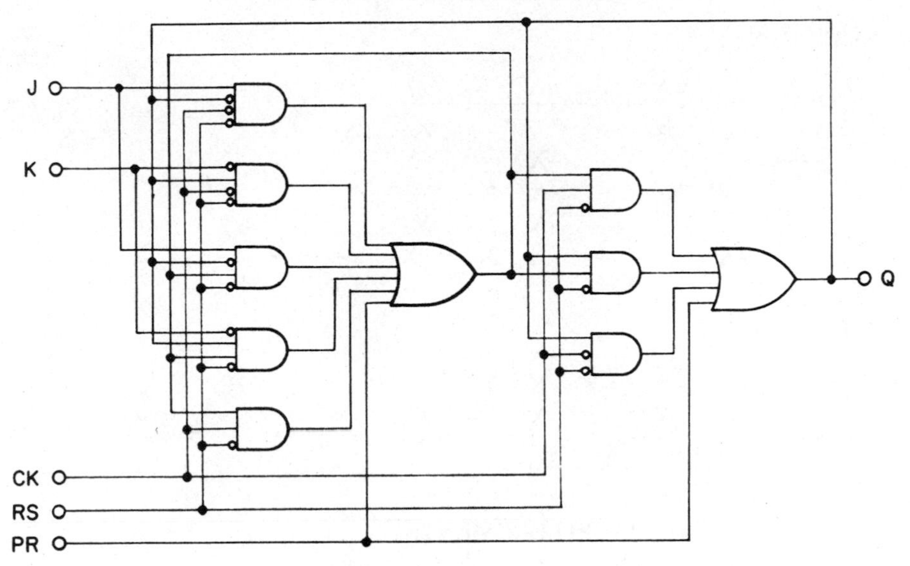

Fig. 8.19 J–K flip-flop circuit diagram.

Logic equations:

$$
\begin{aligned}
X =\ & \overline{CK} * J * Q * \overline{RS} \\
& + \overline{CK} * \overline{K} * \overline{Q} * \overline{RS} \\
& + CK * X * \overline{RS} \\
& + J * Q * X * \overline{RS} \\
& + \overline{K} * \overline{Q} * X * \overline{RS} \\
& + PR
\end{aligned}
$$

$$
\begin{aligned}
Q =\ & CK * X * \overline{RS} \\
& + X * Q * \overline{RS} \\
& + \overline{CK} * Q * \overline{RS} \\
& + PR
\end{aligned}
$$

Truth table:

					Bidirectional Pins			
					Inputs		*Outputs*	
J	*K*	*CK*	*PR*	*RS*	*X*	*Q*	*X*	*Q*
				Active Level			*H*	*H*
H	–	L	–	L	–	H	A	.
–	L	L	–	L	–	L	A	.
–	–	H	–	L	H	–	A	A
H	–	–	–	L	H	H	A	.
–	L	–	–	L	H	L	A	.
–	–	–	H	–	–	–	A	A
–	–	–	–	L	H	H	.	A
–	–	L	–	L	–	H	.	A

X-chart (PAL*m*H*n*):

J	*K*	*CK*	*PR*	*RS*	*X*	*Q*	
X–	--	–X	--	–X	--	X–	(X)
--	–X	–X	--	–X	--	–X	
--	--	X–	--	–X	X–	--	
X–	--	--	--	–X	X–	X–	
--	–X	--	--	–X	X–	–X	
--	--	--	X–	--	--	--	
.							
--	--	X–	--	–X	X–	--	(Q)
--	--	--	--	–X	X–	X–	
--	--	–X	--	–X	--	X–	
--	--	--	X–	--	--	--	

8.2 SEQUENTIAL MACRO ELEMENTS

8.2.1 Introduction

8.2.1.1 Standard registered circuits

Examining the list of standard TTL and CMOS functions reveals that the only standard functions using flip-flops are registers and counters. However, some of the functions which we have described as combinational can usefully be implemented in registered PLDs. One reason for doing this is to synchronise the output with the system clock, in which case the function may be registered by substituting the equivalent registered device for the combinational device, for example, using a PAL16R8 instead of a PAL16L8.

On the other hand, many of the registered parts have a more complex structure than the combinational devices. For example, the PAL16X4 with its exclusive-OR gates or the PLSs with their complement term may be more efficient at containing arithmetic type functions than the basic AND–OR arrays. We will, therefore, look at some of these structures again to see how they fit into the more complex parts.

8.2.2 Registered combinational functions

8.2.2.1 Addressable register (X–PAL)

Logic equations:

$$\overline{Q3} := A1 * A0 * \overline{D} * Q3 * E + A1 * A0 * D * \overline{Q3} * E$$
$$:+: \overline{Q3}$$
$$\overline{Q2} := A1 * \overline{A0} * \overline{D} * Q2 * E + A1 * \overline{A0} * D * \overline{Q2} * E$$
$$:+: \overline{Q2}$$
$$\overline{Q1} := \overline{A1} * A0 * \overline{D} * Q1 * E + \overline{A1} * A0 * D * \overline{Q1} * E$$
$$:+: \overline{Q1}$$
$$\overline{Q0} := \overline{A1} * \overline{A0} * \overline{D} * Q0 * E + \overline{A1} * \overline{A0} * D * \overline{Q0} * E$$
$$:+: \overline{Q0}$$

E	D	A1	A0	Q3	Q2	Q1	Q0
L	X	X	X	q3	q2	q1	q0
H	H	H	H	H	q2	q1	q0
H	L	H	H	L	q2	q1	q0
H	H	H	L	q3	H	q1	q0
H	L	H	L	q3	L	q1	q0
H	H	L	H	q3	q2	H	q0
H	L	L	H	q3	q2	L	q0
H	H	L	L	q3	q2	q1	H
H	L	L	L	q3	q2	q1	L

	E	0	0	0	0	1	1	1	1
	D	0	0	1	1	1	1	0	0
	Q3	0	1	1	0	0	1	1	0
A1	A0								
0	0		H	H			H	H	
0	1		H	H			H	H	
1	1		H	H		H	H		
1	0		H	H			H	H	

Fig. 8.20 4-bit addressable register Karnaugh map.

X-chart:

$\overline{Q3}$	*A1*	$\overline{Q2}$	*A0*	$\overline{Q1}$	*D*	*E*	$\overline{Q0}$	
-X	X-	--	X-	--	-X	X-	--	....
X-	X-	--	X-	--	X-	X-	--	....
X-	--	--	--	--	--	--	--	....
XX	XX	XX	XX	XX	XX	XX	XX	....
--	X-	-X	-X	--	-X	X-	--	....
--	X-	X-	-X	--	X-	X-	--	....
--	--	X-	--	--	--	--	--	....
XX	XX	XX	XX	XX	XX	XX	XX	....
--	-X	--	X-	-X	-X	X-	--	....
--	-X	--	X-	X-	X-	X-	--	....
--	--	--	--	X-	--	--	--	....
XX	XX	XX	XX	XX	XX	XX	XX	....
--	-X	--	-X	--	-X	X-	-X	....
--	-X	--	-X	--	X-	X-	X-	....
--	--	--	--	--	--	--	X-	....
XX	XX	XX	XX	XX	XX	XX	XX	....

	E	0	0	0	0	1	1	1	1
	D	0	0	1	1	1	1	0	0
	Q3	0	1	1	0	0	1	1	0
A1	A0								
0	0								
0	1								
1	1					T		T	
1	0								

Q3

Fig. 8.21 4-bit addressable register Karnaugh maps.

8.2.2.2 Addressable register (PLS)

Logic equations:

$$\begin{aligned}
Q3\!: J = K &= E * A1 * A0 * D * \overline{Q3} \\
&\quad + E * A1 * A0 * \overline{D} * Q3 \\
Q2\!: J = K &= E * A1 * \overline{A0} * D * \overline{Q2} \\
&\quad + E * A1 * \overline{A0} * \overline{D} * Q2 \\
Q1\!: J = K &= E * \overline{A1} * A0 * D * \overline{Q1} \\
&\quad + E * \overline{A1} * A0 * \overline{D} * Q1 \\
Q0\!: J = K &= E * \overline{A1} * \overline{A0} * D * \overline{Q0} \\
&\quad + E * \overline{A1} * \overline{A0} * \overline{D} * Q0
\end{aligned}$$

Truth table:

Inputs				*Present State*				*Next State*			
A1	*A0*	*D*	*E*	*Q3*	*Q2*	*Q1*	*Q0*	*Q3*	*Q2*	*Q1*	*Q0*
H	H	H	H	L	–	–	–	0	–	–	–
H	H	L	H	H	–	–	–	0	–	–	–
H	L	H	H	–	L	–	–	–	0	–	–
H	L	L	H	–	H	–	–	–	0	–	–
L	H	H	H	–	–	L	–	–	–	0	–
L	H	L	H	–	–	H	–	–	–	0	–
L	L	H	H	–	–	–	L	–	–	–	0
L	L	L	H	–	–	–	H	–	–	–	0

8.2.2.3 Registered magnitude comparator

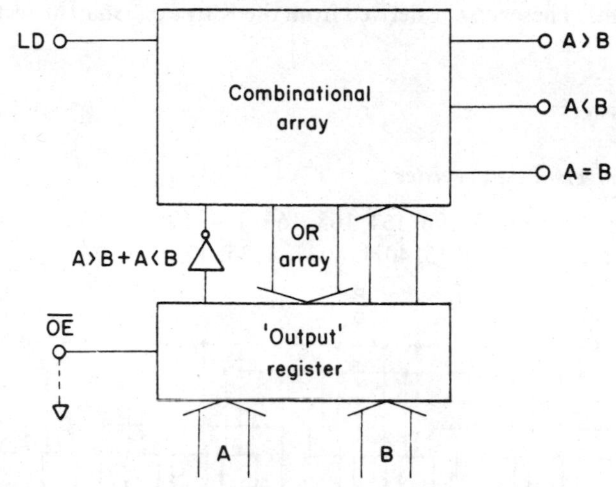

Fig. 8.22 4-bit registered magnitude comparator block diagram.

Description:

The logic equations for this function were detailed in Section 6.2.6.1. By using a PLS159, the inputs can be loaded into a register and the result of comparison obtained from the combinational outputs. Using the complement term to feed back A > B and A < B to generate A = B saves an output, leaving three inputs and one output free for extra logic functions.

Truth table:

			Present State								*Outputs*		
Term	*CT*	*LD*	*A3*	*A2*	*A1*	*A0*	B3	B2	B1	B0	*A>B*	*A<B*	*A=B*
00	A	–	H	–	–	–	L	–	–	–	A		
01	A	–	–	H	–	–	L	L	–	–	A	.	.
.													
.													
12	A	–	H	L	–	H	H	L	L	L	A	.	.
13	A	–	L	–	–	–	H	–	–	–	.	A	.
14	A	–	L	L	–	–	–	H	–	–	.	A	.
.													
.													
25	A	–	H	L	L	L	H	L	–	H	.	A	.
26	.	–	–	–	–	–	–	–	–	–	.	.	A
.													
.													
LB	–	H	–	–	–	–	–	–	–	–			
LA	–	H	–	–	–	–	–	–	–	–			
D3	–	–	–	–	–	–	–	–	–	–			
D2	–	–	–	–	–	–	–	–	–	–			
D1	–	–	–	–	–	–	–	–	–	–			

Note: Only part of the truth table is listed as A > B and A < B each require thirteen terms. These may be derived from the Karnaugh map in Section 6.2.6.1.

8.2.3 Registers

8.2.3.1 Universal shift register

74-series types: 91, 94, 95, 96, 164, 165, 166, 194, 195, 199, 295, 395
4000-series types: 4014, 4015, 4021, 4034, 4035, 4094, 40104

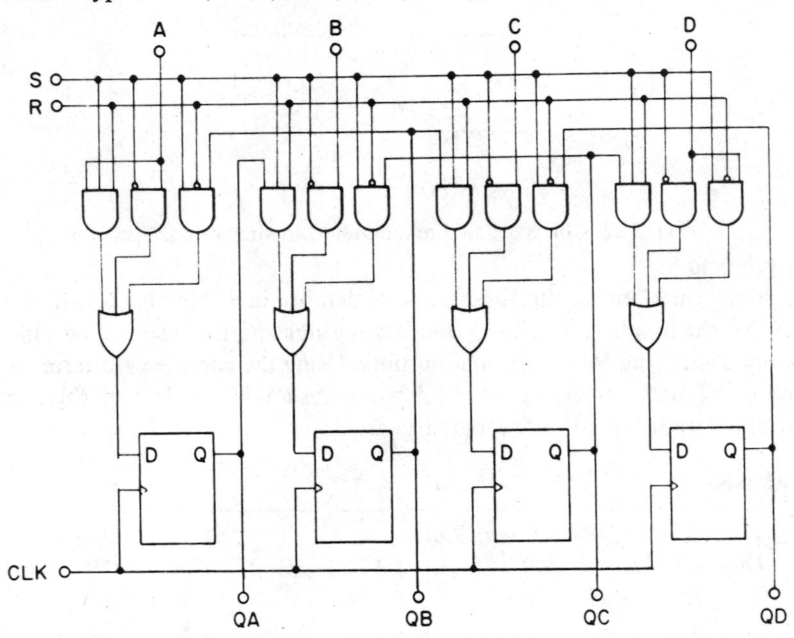

Fig. 8.23 4-bit universal shift register circuit diagram.

Logic equations:

$QA := \overline{S} * A$ (load)
$+ R * S * A$ (shift right)
$+ \overline{R} * S * QB$ (shift left)

$QB := \overline{S} * B$
$+ R * S * QA$
$+ \overline{R} * S * QC$

$QC := \overline{S} * C$
$+ R * S * QB$
$+ \overline{R} * S * QD$

$QD := \overline{S} * D$
$+ R * S * QC$
$+ \overline{R} * S * D$

Truth table (D-type flip-flops):

		Inputs				*Present State*				*Next State*			
R	*S*	*A*	*B*	*C*	*D*	*QA*	*QB*	*QC*	*QD*	*QA*	*QB*	*QC*	*QD*
–	L	H	–	–	–	–	–	–	–	H	–	–	–
H	H	H	–	–	–	–	–	–	–	H	–	–	–
L	H	–	–	–	–	–	H	–	–	H	–	–	–
–	L	–	H	–	–	–	–	–	–	–	H	–	–
H	H	–	–	–	–	H	–	–	–	–	H	–	–
L	H	–	–	–	–	–	–	H	–	–	H	–	–
–	L	–	–	H	–	–	–	–	–	–	–	H	–
H	H	–	–	–	–	–	H	–	–	–	–	H	–
L	H	–	–	–	–	–	–	–	H	–	–	H	–
–	L	–	–	–	H	–	–	–	–	–	–	–	H
H	H	–	–	–	–	–	–	H	–	–	–	–	H
L	H	–	–	–	H	–	–	–	–	–	–	–	H

X-chart (Active-LOW PAL):

R	*S*	*QA*	*A*	*QB*	*B*	*QC*	*C*	*QD*	*D*		
––	–X	––	–X	––	––	––	––	––	––	....	(QA)
X–	X–	––	–X	––	––	––	––	––	––	....	
–X	X–	––	––	X–	––	––	––	––	––	....	
.											
.											
––	–X	––	––	––	–X	––	––	––	––	....	(QB)
X–	X–	X–	––	––	––	––	––	––	––	....	
–X	X–	––	––	––	––	X–	––	––	––	....	
.											
.											
–X	––	––	––	––	––	–X	––	––	....	(QC)	
X–	X–	––	––	X–	––	––	––	––	––	....	
–X	X–	––	––	––	––	––	––	X–	––	....	
.											
	.										
––	–X	––	––	––	––	––	––	–X	....	(QD)	
X–	X–	––	––	––	––	X–	––	––	––	....	
–X	X–	––	––	––	––	––	––	––	–X	....	

8.2.3.2 Registered Barrel Shifter

74 series types: 350 (similar but unregistered)

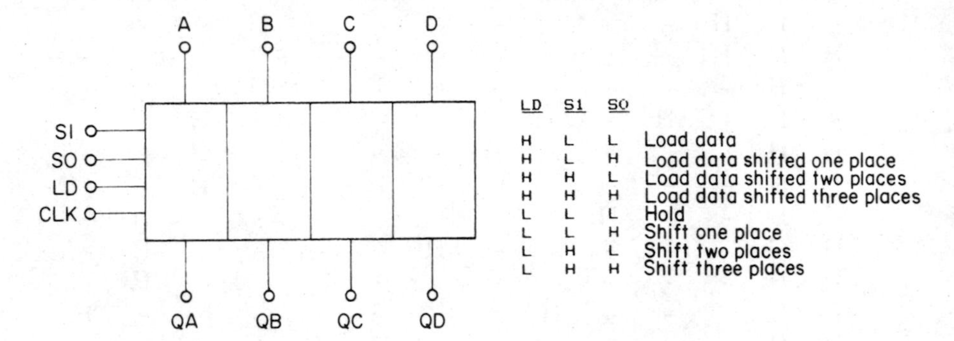

Fig. 8.24 4-bit barrel shifter block diagram and function table.

Logic equations:

$$
\begin{aligned}
QA := {} & LD * \overline{A} * \overline{S1} * \overline{S0} \\
& + LD * \overline{D} * \overline{S1} * S0 \\
& + LD * \overline{C} * S1 * \overline{S0} \\
& + LD * \overline{B} * S1 * S0 \\
& + \overline{LD} * \overline{QA} * \overline{S1} * \overline{S0} \\
& + \overline{LD} * \overline{QD} * \overline{S1} * S0 \\
& + \overline{LD} * \overline{QC} * S1 * \overline{S0} \\
& + \overline{LD} * \overline{QB} * S1 * S0
\end{aligned}
$$

The equations for QB, QC and QD follow a similar pattern.

Truth table (D-type flip-flops):

Inputs							*Present State*				*Next State*			
LD	*S1*	*S0*	*A*	*B*	*C*	*D*	*QA*	*QB*	*QC*	*QD*	*QA*	*QB*	*QC*	*QD*
H	L	L	L	–	–	–	–	–	–	–	A	.	.	.
H	L	H	–	–	–	L	–	–	–	–	A	.	.	.
H	H	L	–	–	L	–	–	–	–	–	A	.	.	.
H	H	H	–	L	–	–	–	–	–	–	A	.	.	.
L	L	L	–	–	–	–	H	–	–	–	A	.	.	.
L	L	H	–	–	–	–	–	–	–	H	A	.	.	.
L	H	L	–	–	–	–	–	–	H	–	A	.	.	.
L	H	H	–	–	–	–	–	H	–	–	A	.	.	.
H	L	L	–	L	–	–	–	–	–	–	.	A	.	.
etc.														

X-chart (Active-LOW PAL):

LD	*S1*	*S0*	*QA*	*A*	*QB*	*B*	*QC*	*C*	*QD*	*D*	
X-	-X	-X	--	-X	--	--	--	--	--	--	....
X-	-X	X-	--	--	--	--	--	--	--	-X	....
X-	X-	-X	--	--	--	--	--	-X	--	--	....
X-	X-	X-	--	--	--	-X	--	--	--	--	....
-X	-X	-X	X-	--	--	--	--	-	--	--	....
-X	-X	X-	--	--	--	--	--	--	X-	--	....
-X	X-	-X	--	--	--	--	X-	--	--	--	....
-X	X-	X-	--	--	X-	--	--	--	--	--	....

8.2.4 Counters

8.2.4.1 Ripple counter

74 series types: 90, 92, 93, 197, 290, 293, 390, 393, 490

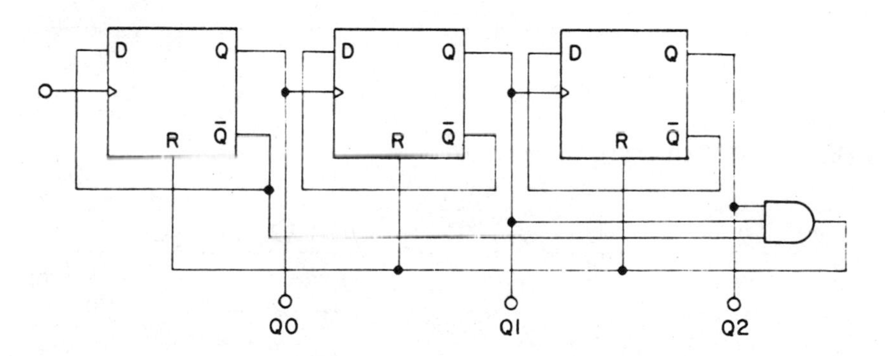

Fig. 8.25 Divide-by-5 ripple counter circuit diagram.

Logic Equations:

$$\begin{aligned} X0 &= \overline{CK} * \overline{Q0} \\ &+ X0 * \overline{Q0} \\ &+ X0 * CK \\ &+ \overline{Q0} * Q1 * Q2 \text{ (reset at count '6')} \\ \overline{Q0} &= CK * \overline{X0} \\ &+ \overline{Q0} * \overline{X0} \\ &+ \overline{CK} * \overline{Q0} \\ &+ Q0 + \overline{Q1} + \overline{Q2} \text{ (reset)} \end{aligned}$$

Q1 and Q2 follow the same equations with suffix '0' replaced by '1' and '2', and CK by Q0 and Q1 respectively.

Truth table:

							Bidirectional Pins					
			Inputs						*Outputs*			
CK	*X0*	*Q0*	*X1*	*Q1*	*X2*	*Q2*	*X0*	*Q0*	*X1*	*Q1*	*X2*	*Q2*
			Active Level				*L*	*L*	*L*	*L*	*L*	*L*
L	–	L	–	–	–	–	–	A	A	.	.	.
–	H	L	–	–	–	–	A	.	.	.	.	.
H	H	–	–	–	–	–	A	.	.	.	.	.
H	L	–	–	–	–	–	.	A	.	.	.	.
	L	L	–	–	–	–	.	A	.	.	.	.
–	–	L	–	L	–	–	.	.	A	A	.	.
–	–	–	H	L	–	–	.	.	A	.	.	.
.												
.												
.												
–	–	L	–	H	–	H	A	.	A	.	A	.
–	–	H	–	–	–	–	.	A	.	A	.	A
–	–	–	–	L	–	–	.	A	.	A	.	A
–	–	–	–	–	–	L	.	A	.	A	.	A

X-chart (PAL*m*L*n*):

CK	*X0*	*I1*	*Q0*	*I2*	*X1*	*I3*	*Q1*	*I4*	*X2*	*I5*	*Q2*	
--	--	--	--	--	--	--	--	--	--	--	--	(TS term)
-X	--	--	-X	--	--	--	--	--	--	--	--	(X0)
--	X-	--	-X	--	--	--	--	--	--	--	--	
X-	X-	--	--	--	--	--	--	--	--	--	--	
--	--	-	-X	--	--	--	X-	--	--	--	X-	
.												
.												
--	--	--	--	--	--	--	--	--	--	--	--	(TS term)
-X	--	--	-X	--	--	--	--	--	--	--	--	(Q0)
X-	-X	--	--	--	--	--	--	--	--	--	--	
--	-X	--	-X	--	--	--	--	--	--	--	--	
--	--	--	X-	--	--	--	--	--	--	--	--	
--	--	--	--	--	--	--	-X	--	--	--	--	
--	--	--	--	--	--	--	--	--	--	--	-X	
.												
.												

If a different count is required, all that is necessary is to modify the reset term; no reset is required if the count is divide-by-2^n.

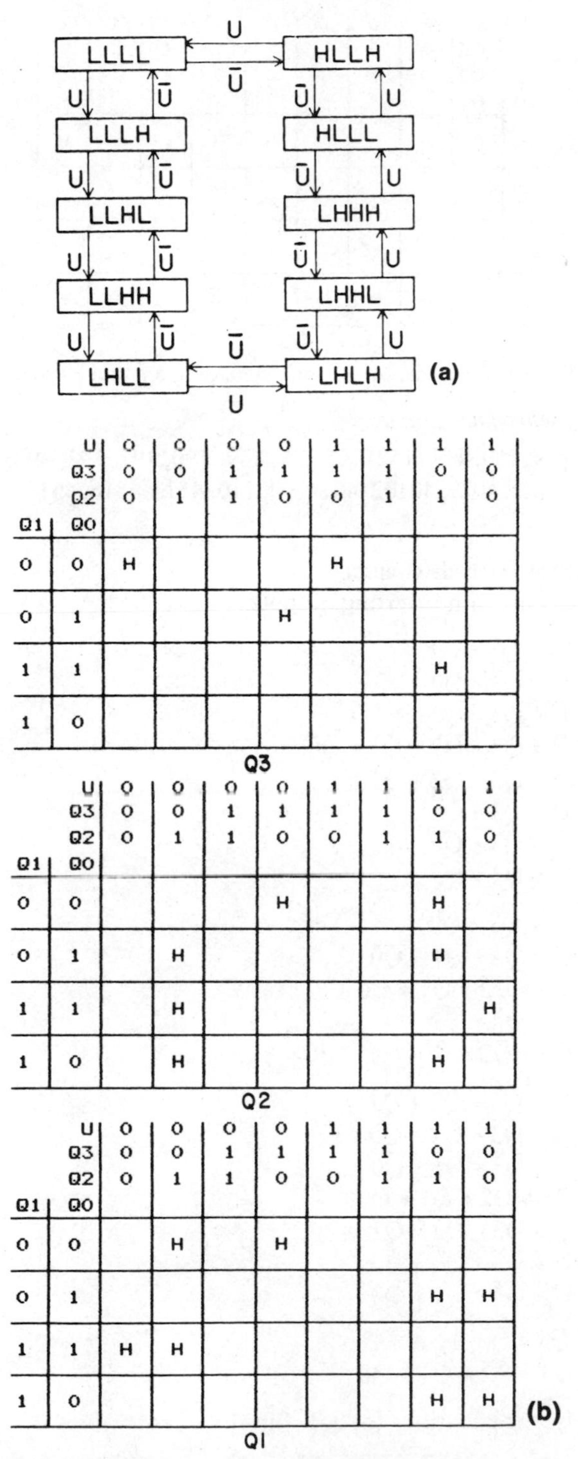

Fig. 8.26 Up/down decade counter state diagram and Karnaugh maps.

Q1	U	0	0	0	0	1	1	1	1
	Q3	0	0	1	1	1	1	0	0
	Q2	0	1	1	0	0	1	1	0
	Q0								
0	0	H	H		H	H		H	H
0	1								
1	1								
1	0	H	H					H	H

Q0 (b)

Fig. 8.26 Up/down decade counter state diagram and Karnaugh maps. (*Cont.*)

8.2.4.2 Synchronous Counters

74-series types: 160, 161, 162, 163,168, 169, 190, 191, 192, 193, 568, 569
4000-series types: 4029, 40102, 40103, 40110, 4510, 4516, 4518, 4520, 4521, 4522, 4526

Logic equations (Decade counter):
(D-type flip-flops with inverting outputs are assumed)

$$\begin{aligned}\overline{Q3} := {} & \overline{U} * \overline{Q3} * Q1 \\ & + \overline{U} * \overline{Q3} * Q0 \\ & + \overline{U} * \overline{Q3} * Q2 \\ & + \overline{U} * Q3 * \overline{Q2} * \overline{Q1} * \overline{Q0} \\ & + U * Q3 * \overline{Q2} * \overline{Q1} * Q0 \\ & + U * \overline{Q3} * \overline{Q1} \\ & + U * \overline{Q3} * \overline{Q0} \\ & + U * \overline{Q3} * \overline{Q2}\end{aligned}$$

$$\begin{aligned}\overline{Q2} := {} & \overline{U} * \overline{Q3} * \overline{Q2} \\ & + \overline{U} * \overline{Q3} * \overline{Q1} * \overline{Q0} \\ & + Q3 * \overline{Q2} * \overline{Q1} * Q0 \\ & + U * \overline{Q2} * \overline{Q1} \\ & + \overline{Q3} * \overline{Q2} * \overline{Q0}\end{aligned}$$

$$\begin{aligned}\overline{Q1} := {} & \overline{U} * \overline{Q3} * \overline{Q2} * \overline{Q1} \\ & + \overline{U} * \overline{Q3} * \overline{Q1} * Q0 \\ & + \overline{U} * \overline{Q3} * Q1 * \overline{Q0} \\ & + Q3 * \overline{Q2} * \overline{Q1} * Q0 \\ & + U * Q3 * \overline{Q2} * \overline{Q1} \\ & + U * \overline{Q3} * \overline{Q1} * \overline{Q0} \\ & + U * \overline{Q3} * Q1 * Q0\end{aligned}$$

$$\begin{aligned}\overline{Q0} := {} & \overline{Q3} * Q0 \\ & + Q3 * \overline{Q2} * \overline{Q1} * Q0\end{aligned}$$

Alternatively, the equations for J–K flip-flops are:

$$\begin{aligned}Q3\text{: } J = K = {} & \overline{U} * \overline{Q2} * \overline{Q1} * \overline{Q0} \\ & + \overline{U} * Q3 * \overline{Q2} * \overline{Q1} * \overline{Q0} \\ & + U * \overline{Q3} * Q2 * Q1 * Q0\end{aligned}$$

$$
\begin{aligned}
\text{Q2: J} = \text{K} &= \overline{\text{U}} * \overline{\text{Q3}} * \text{Q2} * \overline{\text{Q1}} * \overline{\text{Q0}} \\
&+ \overline{\text{U}} * \text{Q3} * \overline{\text{Q2}} * \overline{\text{Q1}} * \overline{\text{Q0}} \\
&+ \text{U} * \overline{\text{Q3}} * \text{Q1} * \text{Q0}
\end{aligned}
$$

$$
\begin{aligned}
\text{Q1: J} = \text{K} &= \overline{\text{U}} * \overline{\text{Q3}} * \text{Q2} * \overline{\text{Q1}} * \overline{\text{Q0}} \\
&+ \overline{\text{U}} * \text{Q3} * \overline{\text{Q2}} * \overline{\text{Q1}} * \overline{\text{Q0}} \\
&+ \overline{\text{U}} * \overline{\text{Q3}} * \text{Q1} * \overline{\text{Q0}} \\
&+ \text{U} * \overline{\text{Q3}} * \text{Q0}
\end{aligned}
$$

$$
\begin{aligned}
\text{Q0: J} = \text{K} &= \overline{\text{Q3}} \\
&+ \text{Q3} * \overline{\text{Q2}} * \overline{\text{Q1}}
\end{aligned}
$$

Truth table:

	Present State				*Next State*			
U	$\overline{Q3}$	$\overline{Q2}$	$\overline{Q1}$	$\overline{Q0}$	$\overline{Q3}$	$\overline{Q2}$	$\overline{Q1}$	$\overline{Q0}$
L	H	H	H	H	0	–	–	–
L	L	H	H	H	0	0	0	–
H	L	H	H	L	0	–	–	–
H	H	L	L	L	0	–	–	–
H	H	L	H	H	–	0	0	–
H	H	–	L	L	–	0	–	–
L	H	–	L	H	–	–	0	–
H	H	–	–	L	–	–	0	–
–	H	–	–	–	–	–	–	0
–	L	H	H	–	–	–	–	0

The J–K solution is so much less complex than the D-type (PAL) solution that we will not offer an X-chart illustration. The equations above are, of course, still valid and may be used for this function if required. The equations for a binary counter were given in Section 5.2.1.5.

8.2.4.3 Johnson counters

4000-series types: 4017, 4022

Logic equations:
In the above Karnaugh maps forbidden states have been shown by an 'f'. By including these as an option the equations can be simplified (to those of a shift register). This will work if the circuit starts in a legal state so, if the power-up condition is not known, a jump to a legal state must be included from all the forbidden states. The equations are thus:

$$
\begin{aligned}
\overline{\text{Q3}} &:= \overline{\text{Q2}} \\
&+ \text{Q2} * \overline{\text{Q1}} * \text{Q0} \\
&+ \overline{\text{Q3}} * \text{Q2} * \overline{\text{Q0}}
\end{aligned}
$$

$$
\begin{aligned}
\overline{\text{Q2}} &:= \overline{\text{Q1}} \\
&+ \overline{\text{Q3}} * \text{Q1} * \overline{\text{Q0}} \\
&+ \text{Q3} * \overline{\text{Q2}} * \text{Q1}
\end{aligned}
$$

$$
\begin{aligned}
\overline{\text{Q1}} &:= \overline{\text{Q0}} \\
&+ \text{Q2} * \overline{\text{Q1}} * \text{Q0} \\
&+ \text{Q3} * \overline{\text{Q2}} * \text{Q0}
\end{aligned}
$$

$$
\begin{aligned}
\overline{\text{Q0}} &:= \text{Q3} \\
&+ \overline{\text{Q3}} * \text{Q2} * \overline{\text{Q1}} \\
&+ \overline{\text{Q3}} * \text{Q1} * \overline{\text{Q0}}
\end{aligned}
$$

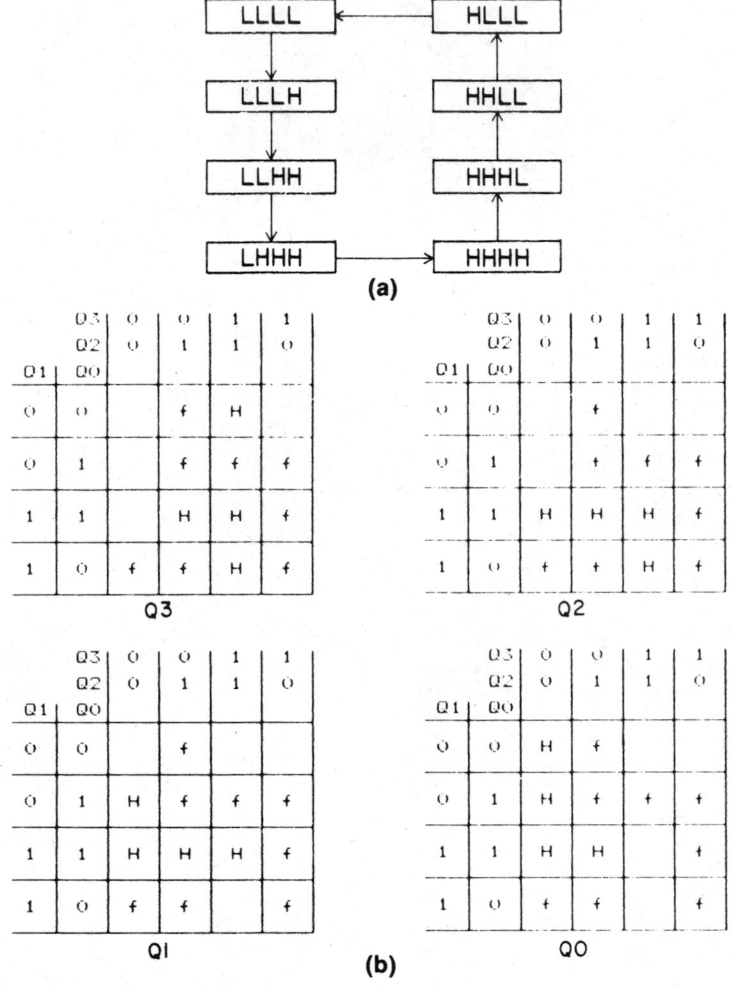

Fig. 8.27 4-bit Johnson counter state diagram and Karnaugh maps.

Truth table:

Thanks to the PLS structure, a more efficient solution may be obtained by including all the jumps from forbidden states together, thus:

Present State				Next State			
$\overline{Q3}$	$\overline{Q2}$	$\overline{Q1}$	$\overline{Q0}$	$\overline{Q3}$	$\overline{Q2}$	$\overline{Q1}$	$\overline{Q0}$
H	L	H	–	H	H	H	H
L	–	H	L	H	H	H	H
L	H	L	–	H	H	H	H
H	–	L	H	H	H	H	H
–	H	–	–	H	–	–	–
–	–	H	–	–	H	–	–
–	–	–	H	–	–	H	–
L	–	–	–	–	–	–	H

X-chart (Active-LOW PAL):

..	$\overline{Q3}$	....	$\overline{Q2}$	....	$\overline{Q1}$	....	$\overline{Q0}$	....	
--	--	--	X-	--	--	--	--	....	($\overline{Q3}$)
--	--	--	-X	--	X-	--	-X	....	
--	X-	--	-X	--	--	--	X-	....	
.									
.									
--	--	--	--	--	X-	--	--	....	($\overline{Q2}$)
--	X-	--	--	--	-X	--	X-	....	
--	-X	--	X-	--	-X	--	--	....	
.									
.									
--	--	--	--	--	--	--	X-	....	($\overline{Q1}$)
--	--	--	-X	--	X-	--	-X	....	
--	-X	--	X-	--	--	--	-X	....	
.									
.									
--	-X	--	--	--	--	--	--	....	($\overline{Q0}$)
--	X-	--	-X	--	X-	--	--	....	
--	X-	--	--	--	-X	--	X-	....	

8.3 MISCELLANEOUS APPLICATIONS

8.3.1 Introduction

This section is devoted to designs which may be incorporated into PLD circuits but which are not generally available as discrete logic circuits in the 74 series or 4000 series. In some cases they may be available as dedicated LSI circuits, although they are mostly too specialised or not complex enough to warrant an independent existence in a manufacturer's catalogue. Even if they do not fit exactly into a design which you want, they are still worth looking at as they may provide ideas or show how various functions can be implemented in PLDs. In general, we list only the logic equations or state table, whichever is more appropriate to the application being described, and the target device.

8.3.2 Specific examples

8.3.2.1 Waveform generator

In many applications it is necessary to generate a complex waveform, or several related waveforms. Examples are:

- processor control
- video controllers
- bubble memory drivers

The waveforms are created from a clock and, sometimes, a synchronisation signal. Figure 8.28 shows a block diagram of the basic system which contains two elements, a counter and a decoder. The waveform period will usually be a fixed number of clock periods, so the counter cycle is continuously dividing by this number, or is reset by the synchronising pulse at the start of each cycle. The decoder defines the signal level for each number of the count.

Almost any PLD, or combination of PLDs, may be used in this application, depending on the number and complexity of the waveforms. For example, the counter could be a divide-by-4096 (ten-bits with a PAL20X10) and the decoder a PLE10P4. This would enable four very complex waveforms to be produced. On the other hand, a PLS155 could produce eight outputs, with a much simpler relationship, but incorporating the counter and decoder in the same device.

As a concrete example let us look at a simple set of signals which might be needed in a video controller. We will see how to produce a frame sync, line sync and picture enable, as shown in Figure 8.29. Frame sync is a half clock frequency signal on lines 0, 1, 254 and 255; line sync is a single pulse at picture element '0' on lines 2–253; picture enable is HIGH on picture elements 20–339 for lines

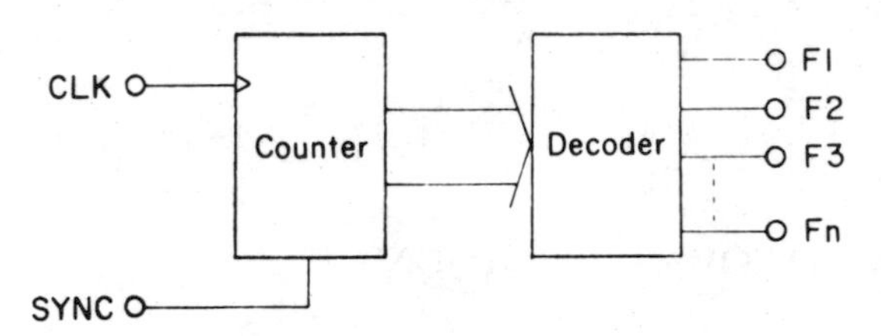

Fig. 8.28 Block diagram of general purpose waveform generator.

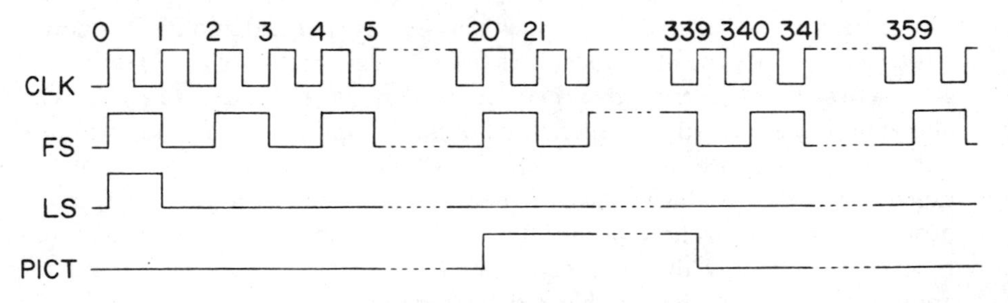

Fig. 8.29 Waveforms for a video controller.

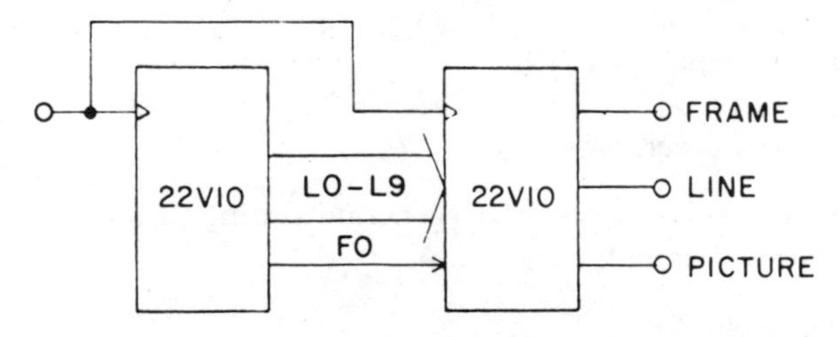

Fig. 8.30 Video controller circuit diagram.

8–247. The entire video raster contains 256 lines of 360 elements, with the picture occupying 240 lines of 320 elements. If the system displays 50 frames per second the picture element rate is 50 × 256 × 360 which is 4.608 MHz, well within the capability of a PLD.

There are many possible ways of building this circuit since the counter requires seventeen flip-flops and the decoder three outputs, decoding seventeen inputs. The minimum chip count solution uses two 22V10s, the first forms a divide-by-360 counter for the line count and also provides the LSB of the frame count. The ten lines are then fed to the other 22V10 which acts as a seven-bit counter to complete the frame, and three outputs are set as combinational to decode the sync pulses. Having already covered counter design at some length, we will just list the equations for the decoder.

$$
\begin{aligned}
\text{LINE} &= \overline{L8} * \overline{L7} * \overline{L6} * \overline{L5} * \overline{L4} * \overline{L3} * \overline{L2} * \overline{L1} * \overline{L0} \\
&\qquad * (\underline{\overline{F7 * F6 * F5 * F4 * F3 * F2 * F1}} \\
&\qquad + \overline{F7} * \overline{F6} * \overline{F5} * \overline{F4} * \overline{F3} * \overline{F2} * \overline{F1}) \\
\text{FRAME} &= F7 * F6 * F5 * F4 * F3 * F2 * F1 * \overline{L1} \\
&+ \overline{F7} * \overline{F6} * \overline{F5} * \overline{F4} * \overline{F3} * \overline{F2} * \overline{F1} * \overline{L1} \\
\overline{\text{PICT}} &= \overline{F7} * \overline{F6} * \overline{F5} * \overline{F4} * \overline{F3} \\
&+ F7 * F6 * F5 * F4 * F3 \\
&+ \overline{L8} * \overline{L7} * \overline{L6} * \overline{L5} * \overline{L4} \\
&+ \overline{L8} * \overline{L7} * \overline{L6} * \overline{L5} * \overline{L4} * \overline{L3} * \overline{L2} \\
&+ L8 * \overline{L7} * L6 * \overline{L5} * L4 * \overline{L3} * L2 \\
&+ L8 * \overline{L7} * L6 * \overline{L5} * L4 * L3 \\
&+ L8 * \overline{L7} * L6 * L5 * \overline{L4} * \overline{L3}
\end{aligned}
$$

F0–F7 and L0–L8 refer to the frame count and line count bits respectively. A circuit diagram of the whole system is shown in Figure 8.30.

8.3.2.2 Code converters and look-up tables

One of the most common 'logic' uses of PROMs/PLEs is as a look-up table, and a particular example is the code converter. We originally described this application in Section 4.1.5.2 and mention it again here for the sake of completeness. Some functions are available as standard parts from various manufacturers; in particular, the BCD-binary conversion, and its converse, are 74S484 and 74S485. Various code conversions such as ASCII–EBCDIC can also be obtained. If the function required is not available as a standard part it will probably be simpler to generate it with a computer program rather than attempt to devise the logic for it.

8.3.2.3 Pseudo random number generator

Figure 8.31 shows the general schematic of a pseudo random number generator. It consists of a shift register with feedback to the input generated as a combinational function of two or more outputs. The register should cycle through every possible combination of states, provided that the function is

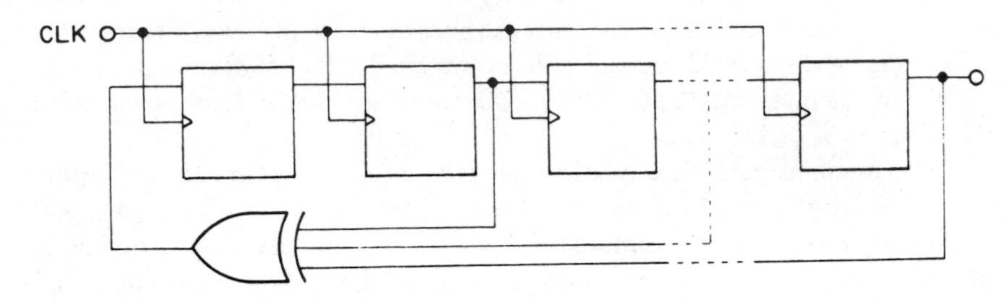

Fig. 8.31 Pseudo random number generator schematic.

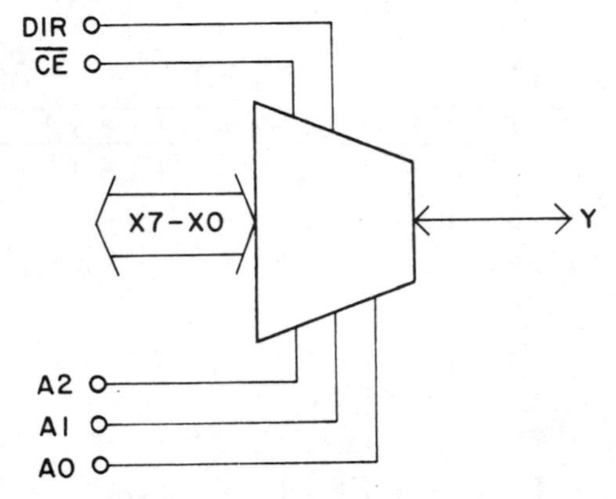

Fig. 8.32 Bidirectional 1-to-8 line mux/demux block diagram.

chosen correctly. Depending how complex the function is, a registered PLE, PAL or PLS can be used to build the 'PRN'. We have already described a standard shift register; the logic array can be used to hold the generator function. For example the function:

Q8 := Q6 :+: Q5 :+: Q4 :+: Q2
Q7 := Q8
Q6 := Q7
Q5 := Q6
Q4 := Q5
Q3 := Q4
Q2 := Q3
Q1 := Q2

gives the following sequence if it starts from FF, once the initial eight HIGHs have worked through:

A8, 54, AA, D5, EA, F5, 7A, 3D, 9E, CF, 67, ...

The only problem with this function is that it cannot start from 00 or 01 as the feedback will always yield a LOW. As with all state machines, care must be taken to ensure that the initial conditions allow correct operation.

8.3.2.3 Bidirectional 1-to-8 line mux/demux

This is an example of a circuit which is a combination of some standard functions described earlier. As such, it could be built from standard TTL or CMOS MSI devices but we will see how these may be replaced by a single PLD. A block diagram is shown in Figure 8.32. If DIR is HIGH then eight-bit wide data (X7–X0) will be multiplexed on to the single data line (Y); if DIR is LOW the single data line will be routed to the parallel port. The address lines, A2–A0, select which of the parallel lines is active, and there is a chip enable, $\overline{\text{CE}}$, which makes every output tri-state when HIGH. The equations for this function are:

$$
\begin{aligned}
&\text{IF}(\overline{\text{CE}} * \text{DIR})\ \text{X7} = \text{A2} * \text{A1} * \text{A0} * \text{Y}\\
&\text{IF}(\overline{\text{CE}} * \text{DIR})\ \text{X6} = \text{A2} * \text{A1} * \overline{\text{A0}} * \text{Y}\\
&\text{IF}(\overline{\text{CE}} * \text{DIR})\ \text{X5} = \text{A2} * \overline{\text{A1}} * \text{A0} * \text{Y}\\
&\text{IF}(\overline{\text{CE}} * \text{DIR})\ \text{X4} = \text{A2} * \overline{\text{A1}} * \overline{\text{A0}} * \text{Y}\\
&\text{IF}(\overline{\text{CE}} * \text{DIR})\ \text{X3} = \overline{\text{A2}} * \text{A1} * \text{A0} * \text{Y}\\
&\text{IF}(\overline{\text{CE}} * \text{DIR})\ \text{X2} = \overline{\text{A2}} * \text{A1} * \overline{\text{A0}} * \text{Y}\\
&\text{IF}(\overline{\text{CE}} * \text{DIR})\ \text{X1} = \overline{\text{A2}} * \overline{\text{A1}} * \text{A0} * \text{Y}\\
&\text{IF}(\overline{\text{CE}} * \text{DIR})\ \text{X0} = \overline{\text{A2}} * \overline{\text{A1}} * \overline{\text{A0}} * \text{Y}\\
&\text{IF}(\overline{\text{CE}} * \overline{\text{DIR}})\ \text{Y} = \text{A2} * \text{A1} * \text{A0} * \text{X7}\\
&\qquad + \text{A2} * \text{A1} * \overline{\text{A0}} * \text{X6}\\
&\qquad + \text{A2} * \overline{\text{A1}} * \text{A0} * \text{X5}\\
&\qquad + \text{A2} * \overline{\text{A1}} * \overline{\text{A0}} * \text{X4}\\
&\qquad + \overline{\text{A2}} * \text{A1} * \text{A0} * \text{X3}\\
&\qquad + \overline{\text{A2}} * \text{A1} * \overline{\text{A0}} * \text{X2}\\
&\qquad + \overline{\text{A2}} * \overline{\text{A1}} * \text{A0} * \text{X1}\\
&\qquad + \overline{\text{A2}} * \overline{\text{A1}} * \overline{\text{A0}} * \text{X0}
\end{aligned}
$$

This function needs nine bidirectional pins with the ability to support eight AND terms on one of them. This is no problem for a PLA such as the PLS153, but it would require a fairly complex PAL, a 22V10 or EP600, for example, to make this function. A PLE solution is out of the question since they do not possess bidirectional pins.

Several variations of the basic function are possible. There are enough inputs to make two completely independent four-line to one-line devices; if a registered device is used, the data can be stored in the output flip-flops. Typical uses for this function are as a simple UART or, for example, to interface to a single-bit device like a dynamic RAM.

8.3.2.5 Crosspoint switch

In a similar vein to the above circuit, the crosspoint switch allows each output to select any input as its data source. The address of the selected input is stored in a latch or register. This design is based on the PLS159 which has enough room for four inputs and four outputs; a larger switch could be accommodated in a larger device, but the equations would be basically the same. An EP900 could support an eight-input/four-output device, for instance, and the PAL64R32 an eight-input/eight-output switch.

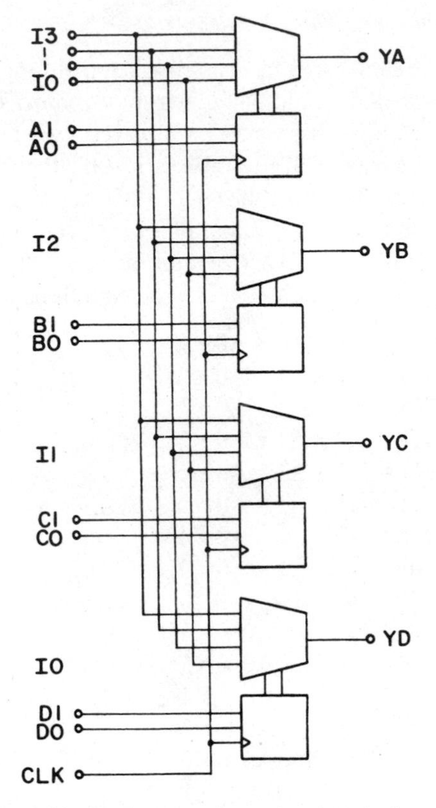

Fig. 8.33 4 × 4 crosspoint switch block diagram.

A block diagram of the device is shown in Figure 8.33 and the equations for the PLS159 are:

$$
\begin{aligned}
YA &= I3 * QA1 * QA0 \\
&\quad + I2 * QA1 * \overline{QA0} \\
&\quad + I1 * \overline{QA1} * QA0 \\
&\quad + I0 * \overline{QA1} * \overline{QA0} \\
YB &= I3 * QB1 * QB0 \\
&\quad + I2 * QB1 * \overline{QB0} \\
&\quad \text{etc.} \\
LA &= LB = 1 \\
D3 &= D2 = D1 = D0 = 1
\end{aligned}
$$

The equations for YB, YC, and YD are the same as those for YA with QB, QC, or QD replacing QA.

8.3.2.6 Majority logic circuit – 1

In some applications it is necessary to make a decision on the basis of whether there are more HIGH inputs than LOWs, or vice versa. Firstly, we will look at

	I5	0	0	0	0	1	1	1	1
	I4	0	0	1	1	1	1	0	0
	I3	0	1	1	0	0	1	1	0
I2	I1								
0	0						H		
0	1			H		H	H	H	
1	1		H	H	H	H	H	H	H
1	0			H		H	H	H	

Fig. 8.34 Karnaugh map – 3 out of 5 majority logic circuit.

a combinational situation where the inputs are presented in parallel. As an example we can assume that there are five inputs, in which case we want the output to be HIGH if three or more inputs are HIGH. The Karnaugh map of this function is shown in Figure 8.34 and it is apparent that ten AND terms are needed to build this circuit from a standard PAL or PLA. The ten terms are those which include any three of the five inputs; mathematically this is calculated by $^5C_3$. This means that the number of terms will increase rapidly as the number of inputs increases.

For more than five inputs, assuming that only odd numbers are considered, it will usually be necessary to use a PLE for this function. The equations for five inputs are:

MAJ = I5 * I4 * I3
+ I5 * I4 * I2
+ I5 * I4 * I1
+ I5 * I3 * I2
+ I5 * I3 * I1
+ I5 * I2 * I1
+ I4 * I3 * I2
+ I4 * I3 * I1
+ I4 * I2 * I1
+ I3 * I2 * I1

8.3.2.7 Majority Logic – 2

If the data are presented in serial form the problem can be tackled as a sequential circuit. A typical application might be to clean up a noisy signal. If '*n*' samples are taken for each majority decision then the circuit must be clocked at a rate $n+1$ times the maximum frequency contained in the signal, or higher. The circuit consists of two counters; one cycles continuously, dividing by 6, the other is only incremented when the signal input is high. On the sixth count the output is set HIGH if the second counter is at '3' or more, otherwise it is set LOW. Both counters are reset by the next clock input and the process repeated. The state diagram for this operation is shown in Figure 8.35. This can be put into either a

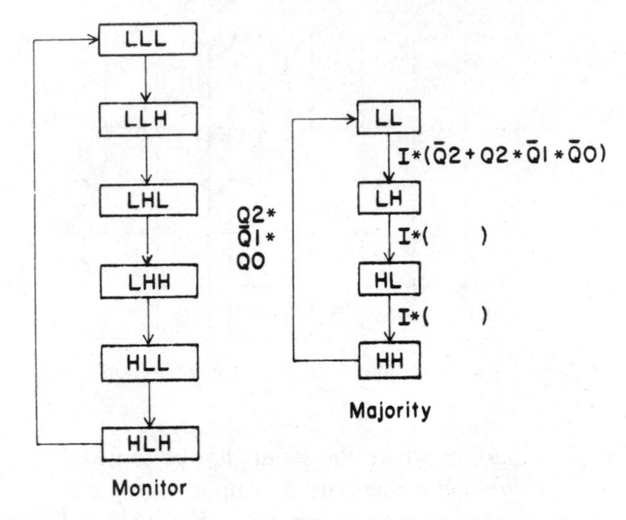

Fig. 8.35 State diagram for sequential 3 out of 5 majority logic circuit.

registered PAL or a PLS. The more complex solution is the PAL, using D-type flip-flops, so we will quote the equations for this method:

Standard counter:

$$\begin{aligned}
\overline{Q2} &:= \overline{Q2} * \overline{Q0} \\
&\quad + Q2 * Q1 \\
&\quad + \overline{Q1} * Q0 \\
\overline{Q1} &:= \overline{Q2} * \overline{Q1} * \overline{Q0} \\
&\quad + \overline{Q2} * Q1 * Q0 \\
&\quad + Q2 \\
\overline{Q0} &:= Q0 \\
&\quad + Q2 * Q1
\end{aligned}$$

Majority counter:

$$\begin{aligned}
\overline{M1} &:= \overline{I} * \overline{M1} \\
&\quad + I * \overline{M1} * \overline{M0} * \overline{(Q2 * \overline{Q1} * Q0)} \\
&\quad + Q2 * \overline{Q1} * Q0 \\
\overline{M0} &:= \overline{I} * \overline{M0} \\
&\quad + I * \overline{M1} * M0 * \overline{(Q2 * \overline{Q1} * Q0)} \\
&\quad + Q2 * \overline{Q1} * Q0
\end{aligned}$$

Output signal:

$$\begin{aligned}
O &:= M1 * M0 * Q2 * \overline{Q1} * Q0 \\
&\quad + O * \overline{(Q2 * \overline{Q1} * Q0)}
\end{aligned}$$

The function '$\overline{Q2 * \overline{Q1} * Q0)}$' is ideally suited to use of the complement term in a PLS. If a device without a complement term is being used the complement may be replaced by '$\overline{Q2} + \overline{Q1} * \overline{Q0}$'.

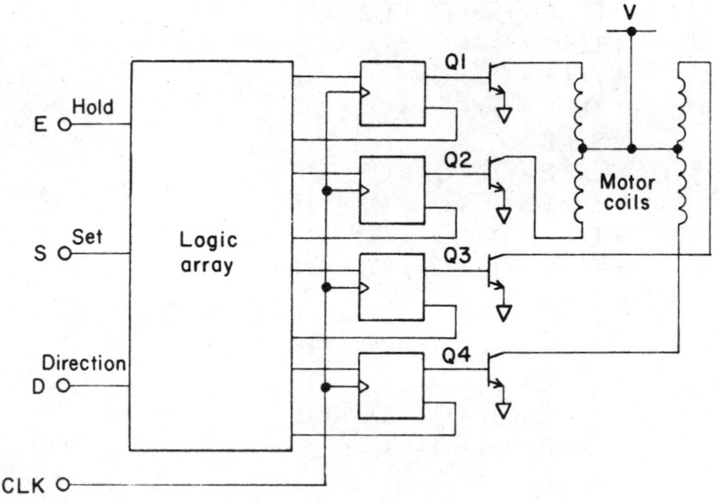

Fig. 8.36 Stepper motor controller block diagram.

8.3.2.8 Stepper motor controller

This application is described in the MMI PAL *Handbook* for both half-step and full-step applications, but the equations are derived for only the full-step case. We will derive the half-step equations. Figure 8.36 shows the block diagram, with Q1–Q4 driving the motor coils, D defining the direction of rotation, E enabling rotation and S the start-up which loads step 1 into the output register. The step sequence is:

STEP	*Q4*	*Q3*	*Q2*	*Q1*
1	0	1	0	1
2	0	0	0	1
3	1	0	0	1
4	1	0	0	0
5	1	0	1	0
6	0	0	1	0
7	0	1	1	0
8	0	1	0	0
1	0	1	0	1

Clockwise rotation is obtained from a sequence 1–2–3–4–5–6–7–8–1, anti-clockwise by 1–8–7–6–5–4–3–2–1. Because the PAL example is based on D-type flip-flops we will illustrate this case with J–Ks. Apart from showing an alternative implementation, this has the advantage that the equations can be derived directly from the step table, which is effectively a state table. We can use toggle mode for the waveform generation while setting step 1 uses direct loading; hold mode is implicit in the J–K. The equations are:

$$
\begin{aligned}
\text{Q4: } J = K = & \overline{D} * E * \overline{Q4} * \overline{Q3} * \overline{Q2} * Q1 \\
& + \overline{D} * E * Q4 * \overline{Q3} * Q2 * \overline{Q1} \\
& + D * E * Q4 * \overline{Q3} * \overline{Q2} * Q1 \\
& + D * E * \overline{Q4} * \overline{Q3} * Q2 * \overline{Q1} \\
K = & S * E \\
\text{Q3: } J = K = & \overline{D} * E * \overline{Q4} * Q3 * \overline{Q2} * Q1 \\
& + \overline{D} * E * \overline{Q4} * \overline{Q3} * Q2 * \overline{Q1} \\
& + D * E * \overline{Q4} * \overline{Q3} * \overline{Q2} * Q1 \\
& + D * E * \overline{Q4} * Q3 * Q2 * \overline{Q1} \\
J = & S * E
\end{aligned}
$$

$$
\begin{aligned}
\text{Q2: } J = K = & \overline{D} * E * Q4 * \overline{Q3} * \overline{Q2} * \overline{Q1} \\
& + \overline{D} * E * \overline{Q4} * Q3 * Q2 * \overline{Q1} \\
& + D * E * \overline{Q4} * Q3 * \overline{Q2} * \overline{Q1} \\
& + D * E * Q4 * \overline{Q3} * Q2 * \overline{Q1} \\
K = & S * E
\end{aligned}
$$

$$
\begin{aligned}
\text{Q1: } J = K = & \overline{D} * E * \overline{Q4} * Q3 * \overline{Q2} * \overline{Q1} \\
& + \overline{D} * E * Q4 * \overline{Q3} * \overline{Q2} * Q1 \\
& + D * E * Q4 * \overline{Q3} * \overline{Q2} * \overline{Q1} \\
& + D * E * \overline{Q4} * Q3 * \overline{Q2} * Q1 \\
J = & S * E
\end{aligned}
$$

An alternative solution could have been obtained by entering the state table directly from the step sequence, instead of generating equations. This method, which is more applicable to R—S-type PLSs, yields an identical number of AND terms.

8.3.2.9 *Shaft encoder*

Also associated with motors is the measurement of shaft speed and direction of rotation. The technique often used is a special case of a quadrature detector. Two signals are generated by the rotating shaft, usually by optical means; these are at the same frequency but shifted in phase by 90°. The speed of rotation is determined by the frequency of the signals and the direction by determining which of the two signals is ahead of the other.

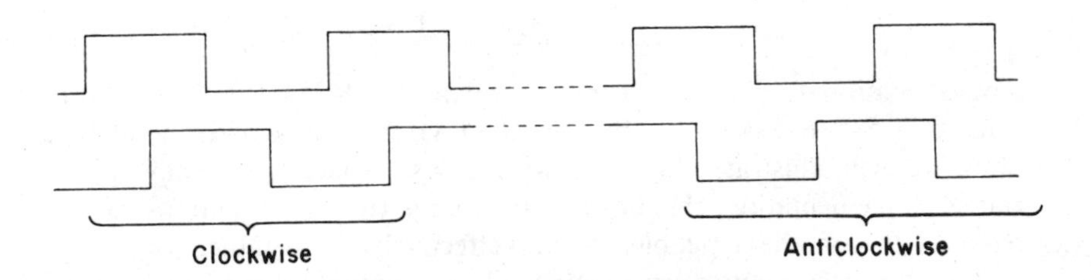

Fig. 8.37 Shaft encoder waveforms.

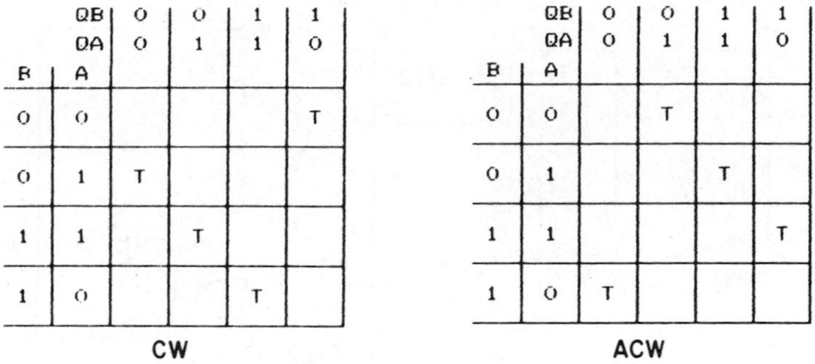

B	A	QB=0, QA=0	QB=0, QA=1	QB=1, QA=1	QB=1, QA=0
0	0				T
0	1	T			
1	1		T		
1	0			T	

CW

B	A	QB=0, QA=0	QB=0, QA=1	QB=1, QA=1	QB=1, QA=0
0	0		T		
0	1			T	
1	1				T
1	0	T			

ACW

Fig. 8.38 Karnaugh maps – shaft encoder.

Figure 8.37 shows the waveforms for clockwise and anti-clockwise rotation. If the waveforms are sampled between the edges of either of them, a sequence of HIGHs and LOWs is obtained which can be analysed with synchronous logic to find the speed and direction. The sequences are:

Clockwise		*Anti-clockwise*	
A	*B*	*A*	*B*
L	L	L	L
H	L	L	H
H	H	H	H
L	H	H	L
L	L	L	L

The waveforms need to be sampled at least four times faster than the maximum signal frequency; the present state of the signals is compared with the previously stored state to obtain the required result. A convenient way of producing the required data is to generate two pulse trains, a clockwise signal which toggles every time an edge is encountered in clockwise rotation, and a similar anti-clockwise signal. Karnaugh maps for driving J–K flip-flops are shown in Figure 8.38, where QA and QB are the stored values of A and B respectively. From these we can derive the following equations:

$$QA := A$$
$$QB := B$$
$$\text{CW: } J = K = \overline{QA} * \overline{QB} * A * \overline{B} + QA * \overline{QB} * A * B + QA * QB * \overline{A} * B + \overline{QA} * QB * \overline{A} * \overline{B}$$
$$\text{ACW: } J = K = \overline{QA} * \overline{QB} * \overline{A} * B + \overline{QA} * QB * A * B + QA * QB * A * \overline{B} + QA * \overline{QB} * \overline{A} * \overline{B}$$

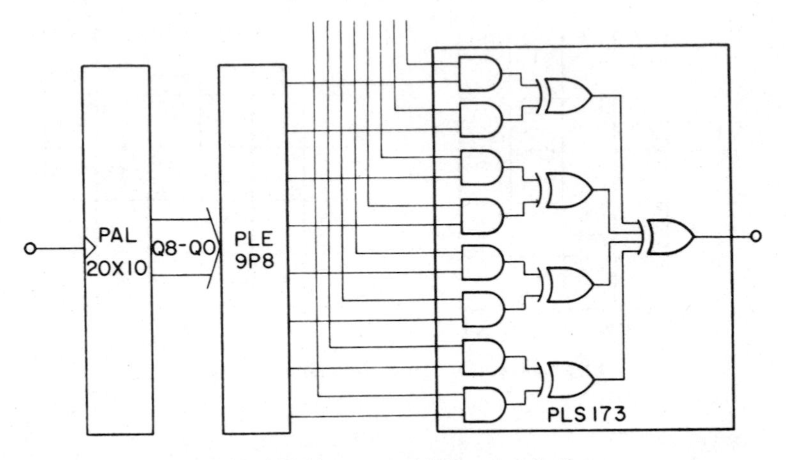

Fig. 8.39 256 bit rate multiplier circuit diagram.

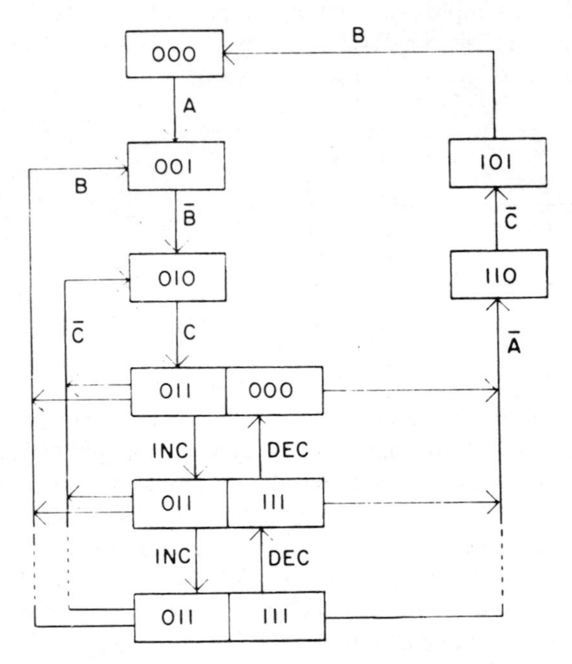

Fig. 8.40 General purpose controller state diagram.

8.3.2.10 Bit rate multiplier

This is a circuit for producing a waveform whose frequency is proportional to the value of some input number entered in binary code. Operation is based on a Gray code counter; this has the property that only one bit changes state at any one time. If two of the outputs are exclusive-ORed then the resulting signal will change state at a rate which is the sum of the two outputs. Referring back to Section 5.3.3.3, we have already seen how to build a Gray code counter. If Q3 toggles at frequency 'f' then Q2 will also toggle at 'f' Q1 at '2f' and Q0 at '4f'. By using the binary inputs to select the appropriate outputs to be exclusive-ORed, any multiple of 'f' from 1 to 7 can be selected. The equation is therefore:

O = S2 * Q0 :+: S1 * Q1 :+: S0 * Q2

As this is an exclusive-OR function the number of AND terms required doubles with each additional bit, unless extensive use is made of feedback terms. A 256-bit rate multiplier would need three PLDs without going to complex devices. Figure 8.39 shows an efficient line-up for this function. A PAL20X10 contains a nine-bit counter, a PLE9P8 is a binary to Gray code converter, and a PLS173 will make an 8-input exclusive-OR gate, configured as the eight AND gates driving four exclusive-OR gates, which feed the final exclusive-OR.

8.3.3 Controllers

8.3.3.1 Control applications

Many systems need a controller to modify their operation or to link components of the system together. The essential features of a controller are that they receive signals from the component parts indicating their status, process these signals and then send further signals to inform the components in what way they should modify their operation.

A simple domestic example is a central heating controller. This takes inputs from temperature sensors in the room and hot-water tank, and from a clock, and sends signals to the pump and boiler depending on whether the heating is needed or not. While this simple example could be satisfied by combinational logic a state machine solution would allow embellishments to be added without disturbing the basic function. As we shall see, more complex systems need synchronous logic for proper operation.

8.3.3.2 A general-purpose controller

Figure 8.40 shows what might be considered to be the state diagram of a general-purpose controller. The controller rest state is '000–000'; on receipt of the start signal 'A' it moves to '001–000' where it waits until 'B' is LOW, thence to '010–000' until 'C' is HIGH. This *soft start* ensures that the process cannot operate until all the components are in the correct state. The initial *run state* is '011–000' which can be modified to '011–001' '011–010', etc. according to the various inputs from the system, thereby adapting operation to the requirements of the system. If, during operation, B or C become invalid the system reverts to its soft-start state until they return to their proper levels.

The system is turned-off by taking 'A' LOW which causes state '110–000' to be entered. When 'C' is LOW it moves to '101–000' until 'B' goes HIGH and the rest state is re-entered. This very generalised picture could be applied to almost any situation where control is required, from gas boilers and arcade games to minicomputers and communication systems via instrumentation and terminals.

In this example, the first three bits form an underlying level of control to the controller itself while the second three bits define the output from the controller to the system. This is ideally suited to a PLS-type of circuit, such as the PLS105 which has a buried register section. This state register can be decoded to give outputs which directly control the system. A typical state table would be:

Inputs					*Present State*						*Next State*					
A	*B*	*C*	*D*	*E*	*F5*	*F4*	*F3*	*F2*	*F1*	*F0*	*F5*	*F4*	*F3*	*F2*	*F1*	*F0*
H	–	–	–	–	L	L	L	L	L	L	L	L	H	L	L	L
H	L	–	–	–	L	L	H	L	L	L	L	H	L	L	L	L
H	L	H	–	–	L	H	L	L	L	L	L	H	H	L	L	L
H	L	H	H	L	L	H	H	L	L	L	L	H	H	L	L	H
H	L	H	H	L	L	H	H	L	L	H	L	H	H	L	H	L
.																
.																
H	L	H	L	H	L	H	H	H	H	H	L	H	H	H	H	L
H	L	H	L	H	L	H	H	H	H	L	L	H	H	H	L	H
.																
.																
H	L	L	–	–	L	H	H	–	–	–	L	H	L	L	L	L
H	H	–	–	–	L	H	–	–	–	–	L	L	H	L	L	L
L	L	H	–	–	L	H	H	–	–	–	H	H	L	L	L	L
L	–	L	–	–	H	H	L	L	L	L	H	L	H	L	L	L
L	H	–	–	–	H	L	H	L	L	L	L	L	L	L	L	L

The inputs ‘D’ and ‘E’ are included as typical system inputs which have the effect of changing the second triplet of state bits in a way which is defined according to the needs of the system. In general, it is better to define state transitions uniquely rather than use ‘don’t cares’ or logic minimisation. From the practical point of view it makes modification much easier, although it may be less efficient in usage of AND terms. From the theoretical point of view it is much safer as there is less risk of opening paths into undefined states or other dead-ends. Checking by Karnaugh map will probably be out of the question because of the large number of inputs which would need to be handled.

8.3.3.3 Handshaking and protocols

In any system there will be a need to exchange data between parts of the system, or with other systems. In parallel with human conversations, apart from the most casual exchanges, protocols exist to determine who should be talking, when a speaker has finished, and who has precedence to take over. In large ‘meetings’ a chairman has to make these decisions, while in smaller conversations the protocol is usually understood by each of the participants. While the rules for human conversation tend to be adaptable, the rules governing electronic data exchange are more rigid. A state machine is often the best way to implement the rules.

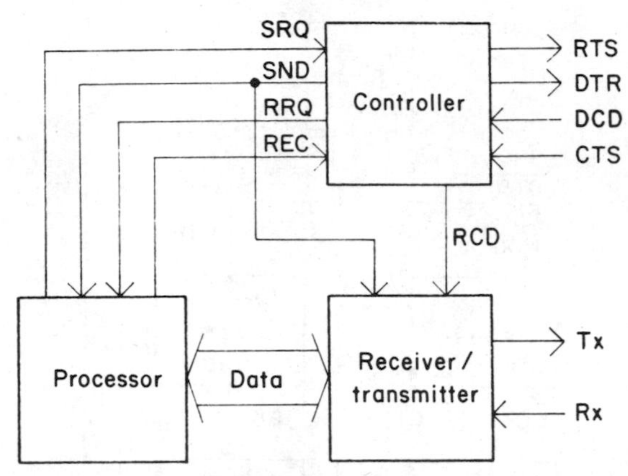

Fig. 8.41 RS232C interface architecture.

As an example we can look at a protocol which is familiar to most users of electronic instruments, RS232C. In order for an instrument to converse via an RS232C link the data flow within the instrument must be managed by an internal controller. The architecture which could be used for an RS232C interface is shown in Figure 8.41. It is assumed that the internal processor can only handle data flow in one direction while the interface has separate transmit and receive lines. The controller must ensure that the interface does not transmit and receive simultaneously, and it must manage the *handshake* with the RS232C lines. That is, it must send an RTS signal when data is to be sent and look for the CTS response; it must also respond to an incoming RTS signal on its DCD input by sending out DTR when the processor is free to receive data.

Internally, the controller is given an SRQ by the processor and sends it SND when all is set up for it to send data. When it receives a request to accept data it sends RRQ to the processor which responds with REC when it is ready; it sends RCD to the transmit/receive circuit at the same time. The state diagram for this is shown in Figure 8.42 and translates into the following state table:

Inputs					*Present State*				*Next State*				
SRQ	*REC*	*DCD*	*CTS*	*SND*	*RRQ*	*RCD*	*RTS*	*DTR*	*SND*	*RRQ*	*RCD*	*RTS*	*DTR*
H	L	L	L	L	L	L	L	L	L	L	L	H	L
H	L	L	H	L	L	L	H	L	H	L	L	H	L
H	L	L	L	H	L	L	H	L	L	L	L	H	L
L	L	L	–	H	L	L	H	L	L	L	L	L	L
–	L	H	L	–	L	L	–	L	–	H	L	–	L
L	H	H	L	–	H	L	–	L	L	H	H	L	H
L	L	H	L	L	L	H	H	L	L	H	L	L	L
–	–	L	L	L	H	–	L	–	L	L	L	L	L

The above table is a simplified version of RS232C sufficient to illustrate the principle of handshaking. In it, the processor is allowed to choose priority

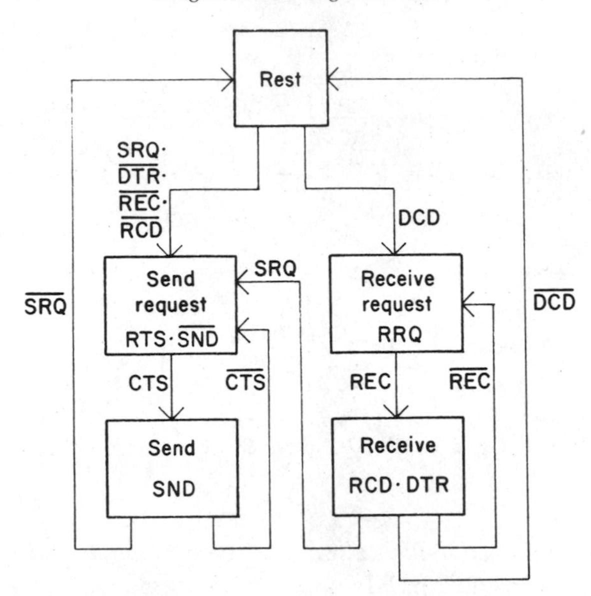

Fig. 8.42 RS232C interface state diagram.

between sending and receiving, although the controller cannot set DTR HIGH unless CTS is LOW.

8.3.3.4 Multi-processor controller

Many systems now contain several processors sharing resources, such as memory and I/O ports, via a common bus. In order to prevent contention and hogging, the bus may be allocated on a time-sharing basis among the processors. This requires a central controller to receive requests for access to the bus and to grant them on a fair basis.

There have been several schemes devised in order to make best use of the resources while allowing a fair distribution of time among the processors. For example, a prioritised system will allow the most important, as judged by the designer, to have first choice even if a lower-priority device is in the middle of an operation. This may be satisfactory for some systems, where the roles of the different processors can be evaluated in terms of their importance. A more egalitarian method is the 'round robin' or Last Granted Lowest Priority (LGLP) structure.

LGLP allows each processor a maximum time to use the bus, after which it checks whether any other processor is requesting access. The last device to use the bus becomes the lowest-priority device, but then moves up the pecking order as each processor has a chance to request access. A PLS is again the ideal device to program for this function. Figure 8.43 shows a PLS in this situation, controlling three processors, and the state diagram required for the operation.

The three processors each have a request line to the controller and a grant line back, which allows access when LOW. When a request is granted the timer is

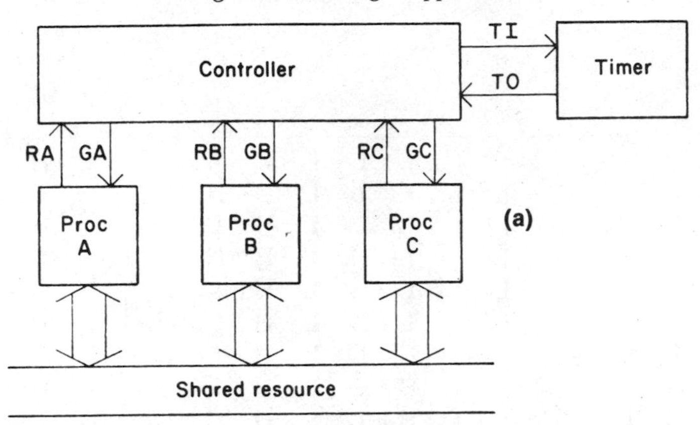

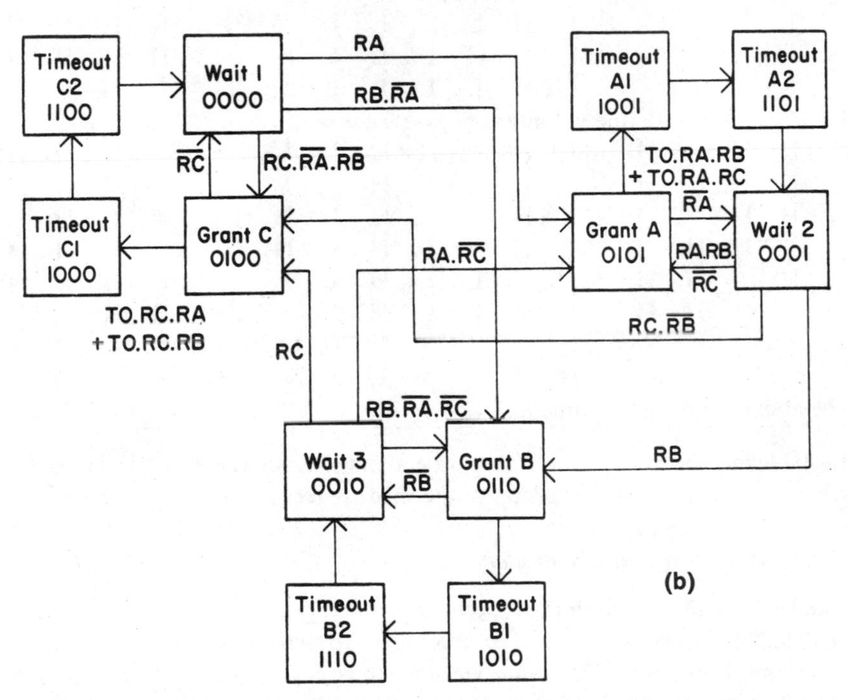

Fig. 8.43 Multi-processor controller. (a) circuit diagram; (b) state diagram.

started and sends back a timeout signal when time is up, unless the processor relinquishes the bus voluntarily. If another processor is requesting at timeout the grant is removed after three clock pulses and the next grant made. There is always a single clock pulse pause between release and a new grant to avoid contention on the bus. The buried register of the PLS may be used to time the internal delays although the timeout is likely to be too long to use the buried register for that. The state table is therefore:

F												Output			
													GB	GC	TI
H	–	–	–	L	L	L	L	L	H	L	H	L	H	H	L
L	H	–	–	L	L	L	L	L	H	H	L	H	L	H	L
L	L	H	–	L	L	L	L	L	H	L	L	H	H	L	L
–	H	–	–	L	L	L	H	L	H	H	L	H	L	H	L
–	L	H	–	L	L	L	H	L	H	L	L	H	H	L	L
H	L	L	–	L	L	L	H	L	H	L	H	L	H	H	L
–	–	H	–	L	L	H	L	L	H	L	L	H	H	L	L
H	–	L	–	L	L	H	L	L	H	L	H	L	H	H	L
L	H	L	–	L	L	H	L	L	H	H	L	H	L	H	L
(The above terms define round robin priority)															
L	–	–	–	L	H	L	H	L	L	L	H	H	H	H	H
–	L	–	–	L	H	H	L	L	L	H	L	H	H	H	H
–	–	L	–	L	H	L	L	L	L	L	L	H	H	H	H
(The above terms define voluntary release)															
H	H	–	H	L	H	L	H	H	L	L	H	L	H	H	H
H	–	H	H	L	H	L	H	H	L	L	H	L	H	H	H
H	H	–	H	L	H	H	L	H	L	H	L	H	L	H	H
–	H	H	H	L	H	H	L	H	L	H	L	H	L	H	H
H	–	H	H	L	H	L	L	H	L	L	L	H	H	L	H
–	H	H	H	L	H	L	L	H	L	L	L	H	H	L	H
–	–	–	–	H	L	–	–	H	H	–	–	–	–	–	–
–	–	–	–	H	H	–	–	L	L	–	–	–	–	–	–
(The above terms define timeout)															

This table may be extended for more processors or converted to a fixed priority, the latter by using only one group of the priority terms.

8.3.3.5 Dual port RAM controller

A dual port RAM is a random access memory which can be written or read from two places. It allows two systems, or processors, to share a common memory and to communicate using the same storage area. Some dedicated devices are available, but these are limited in size and tend to be rather expensive. If it were possible to use a standard memory then more flexible and cheaper arrangements would be possible. What is needed is a controller circuit which will prevent both sides trying to access the RAM simultaneously and ensure a minimum waiting time if one side tries to access the memory while it is already being used.

A state machine is the ideal way of doing this and a PLS will provide the basis for implementing the function in hardware. The circuit for the dual port RAM is shown in Figure 8.44, along with the state diagram. Buffers are needed for the address and data lines, but these would probably be used anyway so do not represent an additional overhead. Typical buffers would be 74LS244 for the addresses and 74LS245 for the data lines.

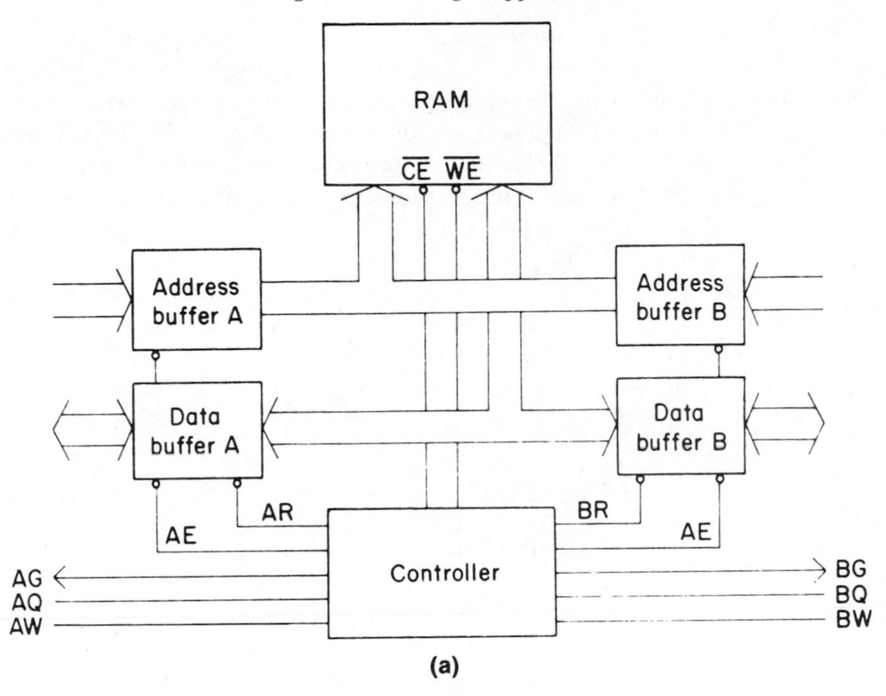

(a)

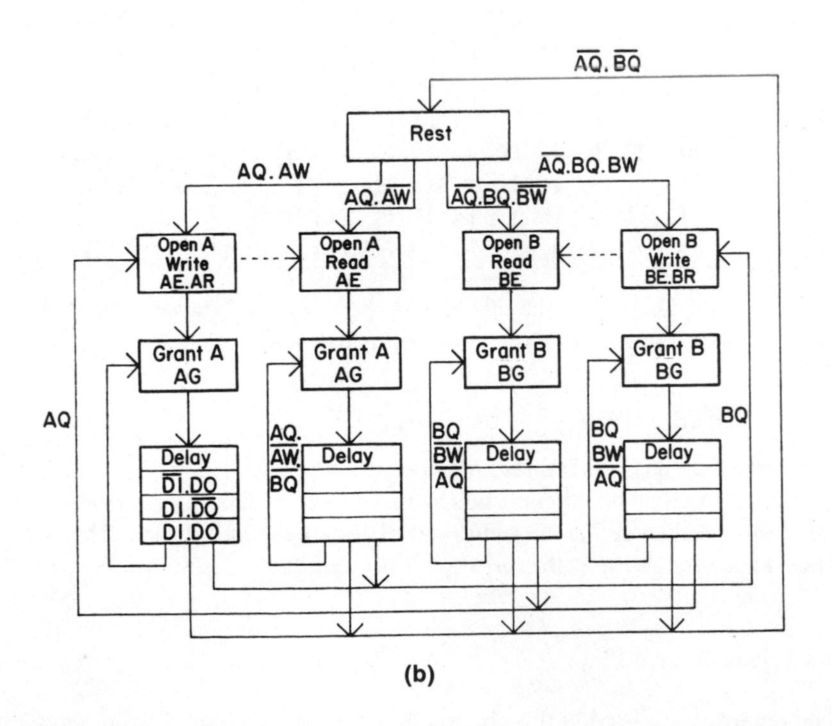

(b)

Fig. 8.44 Dual port RAM controller. (a) circuit diagram; (b) state diagram.

Priority has to be built into the design in case of simultaneous requests for access from A and B sides; in this case we will give priority to A. The state diagram allows for a delay between opening the enabled port and granting access, and between removing the grant and closing the port. This makes sure that the lines are settled before access is attempted. The port is left open for a fixed time in each cycle, and at the end of each cycle control is passed to the opposite side if it is requesting access. The state diagram may be converted to a state table as follows:

Inputs				*Present State*								*Next State*							
AQ	*AW*	*BQ*	*BW*	*AE*	*AR*	*AG*	*BE*	*BR*	*BG*	*D1*	*D0*	*AE*	*AR*	*AG*	*BE*	*BR*	*BG*	*D1*	*D0*
H	H	–	–	L	L	L	L	L	L	L	L	H	H	L	L	L	L	L	L
H	L	–	–	L	L	L	L	L	L	L	L	H	L	L	L	L	L	L	L
–	–	–	–	H	–	L	L	L	L	L	L	H	–	H	L	L	L	L	L
L	–	H	H	L	L	L	L	L	L	L	L	L	L	L	H	H	L	L	L
L	–	H	L	L	L	L	L	L	L	L	L	L	L	L	H	L	L	L	L
–	–	–	–	L	L	L	H	–	L	L	L	L	L	L	H	–	H	L	L
–	–	–	–	–	–	H	–	–	–	L	L	–	–	–	–	–	–	L	H
–	–	–	–	–	–	–	–	–	H	L	L	–	–	–	–	–	–	L	H
–	–	–	–	–	–	–	–	–	–	L	H	–	–	–	–	–	–	H	L
–	–	–	–	–	–	–	–	–	–	H	L	–	–	–	–	–	–	H	H
H	H	L	–	H	H	H	L	L	L	H	H	H	H	H	L	L	L	L	L
H	L	L	–	H	L	H	L	L	L	H	H	H	L	H	L	L	L	L	L
L	–	H	H	L	L	L	H	H	H	H	H	L	L	L	H	H	H	L	L
L	–	H	L	L	L	L	H	L	H	H	H	L	L	L	H	L	H	L	L
H	H	–	–	L	L	L	H	–	H	H	H	H	H	L	L	L	L	L	L
H	L	–	–	L	L	L	H	–	H	H	H	H	L	L	L	L	L	L	L
–	–	H	H	H	–	H	L	L	L	H	H	L	L	L	H	H	L	L	L
–	–	H	L	H	–	H	L	L	L	H	H	L	L	L	H	L	L	L	L
L	–	L	–	–	–	–	–	–	–	H	H	L	L	L	L	L	L	L	L

The control signals for the RAM can be derived from the state bits as combinational functions by the following equations:

$$\mathrm{CE} = \mathrm{AE} + \mathrm{BE}$$
$$\overline{\mathrm{WE}} = \overline{\mathrm{AR}} + \overline{\mathrm{BR}}$$

In the above table the first three lines open the A port and grant A; the next three lines open the B port and grant B. The following five lines implement the delay function and then define the routes out of the final delayed state. The last line takes the system back to the rest state if no requests are pending.

8.3.4 Summary

The applications listed in this chapter are a small selection of what is possible with PLDs. They have been selected to try to cover as many different design areas as possible, but there is clearly scope for many more variations on the basic themes. Some of the designs are original ideas while others have been adapted from ideas generated in manufacturers' handbooks. Most of the PLD manufacturers publish application briefs but in particular the following are recommended for further ideas:

- Signetics Corp. (Philips in Europe)
- Monolithic Memories, Inc.
- Altera, Inc.

Conclusion

This is a conclusion to the book but not to the subject. With a technology developing as fast as programmable logic it is reasonable to try to predict the likely path it will follow. As electronic components progress in size and complexity, it is no good promoting an approach which will become outdated in a short time. ROMs and RAMs have reached the 'megabit' stage, gate arrays can offer complexities in five figures and the thirty-two-bit microprocessor is commonplace, so where is programmable logic going?

The companies that make megabit memories, etc. are also the PLD manufacturers, so there is little chance that technology will leave them behind. Established manufacturers are developing their ideas into larger and more complex architectures, and new companies are springing up with other exciting challenges for the technology experts to make. So what will the PLD market have to offer in ten years' time?

There is already a trend towards CMOS as the technology for building PLDs. We are a little behind other device families in this respect, but even today it can offer performance equal, or superior, to bipolar devices. The limit of complexity has probably been reached with bipolar technology in the mega-PALs and proposed PML architectures. The power dissipation required for larger structures will put them beyond the reach of economic packages. Architecture is moving towards a less formal structure away from the PAL/PLA straitjacket, with its dependence on AND–OR logic, towards a structure with a similarity to gate arrays with freedom of internal interconnections, although a PAL cell may form the basic internal logic unit in such a structure.

The strongest trend is to the use of CAE, which will make the device architecture transparent to the user. The parallel here is with microprocessors, whose development is dependent entirely on a development system. Once a company realises that using PLDs can offer them as much as using microprocessors, a PLD development system will become as common as an MDS.

The cost of SSI components is largely dependent on packaging costs. While a PLD die will never be as cheap as an SSI logic die, the comparative cost could well sink to the point where the less complex discrete logic components are almost obsolete. It may be absurd, at present, to use a simple PAL to replace a single 74LS00, but the equation may not be so obvious in 1996.

The other competing technology is the masked ASIC. These are certainly becoming cheaper, and available in smaller quantities and with shorter lead times. I believe that they will never compete with the ability to create the first sample in a matter of minutes and on the designer's own bench. It is possible that production volumes could turn more to masked devices, such as the HAL, as programming is an annoying extra process for some production areas. Again, as the cost of the silicon becomes a smaller proportion of the device cost, masked devices will start to lose their economic advantage over programmable devices.

Altogether, I can predict only an increasing use of PLDs, both as an absolute number and as a proportion of the total. Anybody starting down the PLD path today will find more, rather than fewer, opportunities to improve his or her performance and to create more efficient logic systems.

Appendix 1

PLD Manufacturer Details

The list below gives the following details of PLD manufacturers:

- addresses in Europe and USA
- ranges of PLDs supplied
- dedicated CAE support
- literature for further reading

ALTERA CORPORATION

UK – Ambar-Cascom Ltd., Rabans Close, Aylesbury, Bucks, HP19 3RS phone: 0296 34141; telex: 837427

France – Tekelec-Airtronic, various addresses

Germany – Electronic 2000, various addresses

USA – 3525 Monroe Street, Santa Clara, California, CA 95051 phone: (408) 984-2800; telex: 888496

Range supplied: EP Series – erasable CMOS technology, medium-complexity and LSI PALs

CAE support: A + PLUS – sophisticated design capture, logic simulation and manipulation and schematic design interfaces; dedicated device programmer

Literature: handbook including application notes

ADVANCED MICRO DEVICES

UK – AMD House, Goldsworth Road, Woking, Surrey, GU21 1JT phone: 04862 22121; telex: 859103

France – Silic 314, Immeuble Helsinki 74, rue d'Arcueil, F-94588 Rungis Cedex phone: 01687 36 66; telex: 202053

Germany – Rosenheimer Strasse 143B, 8000 Munchen 80 phone: 49 89 41140; telex: 05-23883

USA – 901 Thompson Pl., PO Box 3453, Sunnyvale, California, CA 94088 phone: (408) 732-2400; telex: 346306

Range supplied: PLEs medium-complexity PALs, 18P8 and 22V10

CAE support: PLPL – Boolean entry, logic simulation and compilation

Literature: data book

CYPRESS SEMICONDUCTOR CORPORATION

UK – Pronto Electronic Systems Ltd, 466-478 Cranbrook Road, Gants Hill, Ilford, Essex, IG2 6LE phone: 01-554 6222; telex: 8954213

USA – 3901 North First Street, San Jose, California, CA. 95134 phone: (408) 943-2600

Range supplied: erasable CMOS medium-complexity PALs and PLEs

CAE support: none notified

Literature: data sheets

HARRIS CORPORATION

UK – Harris/MHS, Eskdale Road, Winnersh Triangle, Wokingham, Berks, RG11 5TR phone: 0734 698787; telex: 848174

France – Matra-Harris Semiconducteurs, Les Quadrants, 3 Ave du Centre, 78182 Saint Quentin en Yvelines phone: 3-043-82-72; telex: 697317

Germany – Harris/MHS, Erfurterstrasse 29, D-8057 Eching phone: 49-89-319-1035; telex: 524126

USA – PO Box 883, Melbourne, Florida 32901 phone: (305) 724-7000; telex: 510-959-6259

Range supplied: CMOS medium-complexity PALs and PLEs ('zero' quiescent current)

CAE support: HELP – Boolean logic entry and simulation

Literature: Data book and application notes

INTERNATIONAL CMOS TECHNOLOGY, Inc.

UK – Sequoia Technology Ltd, Unit 2, First Avenue, Globe Business Park, Marlow, Bucks, SL7 1YA phone: 06284 76726; telex: 846536

USA – 2031 Concourse Drive, San Jose, California, CA 95131 phone: (408) 434-0678; telex: 910-997-1531

Range supplied: electrially-erasable CMOS PEEL18CV8

CAE support: none notified

Literature: data sheet

INTEL CORPORATION

UK – Piper's Way, Swindon, Wiltshire, SN3 1RJ phone: 0793 696000; telex: 444447

France – 1 Rue Edison, BP303, 78054 Saint Quentin en Yvelines phone: 0130 64 60 00; telex: 69901677

Germany – Seidlstrasse 27, D-8000 Munchen 2 phone: 89 53891; telex: 0523177

USA – 3065 Bowers Ave., Santa Clara, California, CA 95051 phone: (408) 987 8080

Range supplied: erasable CMOS medium-complexity and LSI PALs (second source to Altera)

CAE support: iPLDS – sophisticated logic entry, simulation and compilation and dedicated programmer

Literature: data sheets

LATTICE SEMICONDUCTOR CORPORATION

UK – Macro-Marketing Ltd, Burnham Lane, Slough, SL1 6LN phone: 06286 4422; telex: 847945

USA – 15400 NW Greenbrier Parkway, Beaverton, Oregon, OR 97006 phone: (503) 629-2131; telex: 277338

Range supplied: electrically-erasable CMOS GALs16V8 and 20V8

CAE support: none notified

Literature: data book

MONOLITHIC MEMORIES INCORPORATED

UK – Monolithic House, 1 Queen's Road, Farnborough, Hants, GU14 6DJ phone: 0252 517431; telex: 858051

France – Silic 463, F 94613 Rungis Cedex phone: 1686 08 18; telex: 202146

Germany – Mauerkircherstrasse 4, D 8000 Munchen 80 phone: 89-984961; telex: 524385

USA – 2175 Mission College Boulevard, Santa Clara, California, CA 95050 phone: (408) 970-9700; telex: (910) 338-2374

Range supplied: low-complexity, medium-complexity and LSI bipolar PALs, HALs and PLEs

CAE support: PALASM – Boolean logic entry, simulation and compilation

PLEASM – Boolean logic entry,simulation and compilation for PLEs

Literature: Handbook including Application notes and design examples

NATIONAL SEMICONDUCTOR CORPORATION

UK – 301 Harpur Centre, Horne Lane, Bedford, MK40 1TR phone: 0234 270027; telex: 826209

France – Expansion 10000, 28 Rue de la Redoute, 92260 Fontenay aux Roses phone: 01660 81 40; telex: 250956

Germany – Elsenheimerstrasse 61/11, 8000 Munchen 21 phone: 089 576091; telex: 5222772

USA – 2900 Semiconductor Drive, Santa Clara, California, CA 95051 phone: (408) 737-5000; telex: (910) 339-9240

Range supplied: low-complexity, medium-complexity bipolar PALs, and PLEs

CAE support: none notified

Literature: data sheets

PANATECH/RICOH

Range supplied: erasable CMOS low and medium complexity PALs

SIGNETICS CORPORATION

UK – Mullard Ltd, Torrington Place, London, WC1E 7HD phone: 01-580 6633; telex: 264341

France – RTC La Radiotechnique-Compelec, Ave Ledru-Rollin, Paris phone: 01-335 44 99

Germany – Valvo GmbH, Hamburg phone: 040-3296-19

USA – 811 E. Arques Avenue, PO Box 3409, Sunnyvale, California, CA 94088 phone: (408) 739-7700

Range supplied: bipolar PLAs, PLSs and PLEs

CAE support: AMAZE – Boolean logic entry, simulation and compilation

Literature: data book including application notes

TEXAS INSTRUMENTS INCORPORATED

UK – Manton Lane, Bedford, MK41 7PA phone: 0234 67466; telex: 82178

France – BP5, 06270 Villeneuve Loubet phone: 9320 01 01; telex: 470127

Germany – Haggertystrasse 1, 8050 Freising phone: 08161 800; telex: 526529

USA – PO Box 225012, Dallas, Texas, Tex 75265 phone: (214) 995-6531

Range supplied: low-complexity, medium-complexity bipolar PALs, and PLAs, PLSs and PLEs

CAE support: none notified

Literature: data books

VLSI TECHNOLOGY INCORPORATED

UK – Midsummer House, Midsummer Boulevard, Milton Keynes, MK9 3BN phone: 0908 667595; telex: 825135

France – 40 Rue du Seminaire, Centra 416, F-94616 Rungis Cedex phone: 01687 31 41; telex: 206608

Germany – Rosenkavalierplatz 10, D-8000 Munchen 81 phone: 089 926905-0; telex: 5214279

Range supplied: erasable CMOS medium-complexity PALs with Xpandor cells, and 16V8

CAE support: EPLCONV – converts standard PAL files to VTI compatible files

Literature: data sheets

In general, where no dedicated software is offered, the range is supported by universal programs.

Appendix 2

CAE Manufacturer Details

Listed below are details of manufacturers offering CAE support, together with a summary of the capability of the hardware or software.

DAISY SYSTEMS CORPORATION

UK – Berk House, Basing View, Basingstoke, Hants, RG21 2HQ phone: 0256 464061; telex: 858071

USA – 700 Middlefield Road, Mountain View, California, CA 94039 phone: (415) 960-0123; telex: 858262

Capability: Logician series of work-stations for design entry, simulation and test analysis. PLDMaster converts design data into format suitable for compilation by most universal and dedicated programs (ABEL, CUPL, PALASM, AMAZE, etc.)

DATA I/O – FUTURENET

UK – Microsystem Services, PO Box No. 37, Lincoln Road, Cressex Industrial Estate, High Wycombe, Bucks, HP12 3XJ phone: 0494 41661; telex: 837187

France – M B Electronique, F-78530, Buc phone: 03956 81 31; telex: 695414

Germany – Bahnhofstrasse 3, D-6453 Sligenstadt phone: 06182 3088; telex: 4184962

USA – 10525 Willows Road NE, PO Box 97046, Redmond, Washington, WA 98073 phone: (206) 881-6444; telex: 152167

Capability: 'Personal Silicon Foundry' based on DASH-2 or DASH-3C design entry work-stations. Interface to ABEL universal logic entry, simulation and compilation software, and PLDtest for test analysis. Model 29B programming system is a universal programmer with modules and socket adaptors for all PLDs. Model 60A programmer is capable of programming PALs, PLAs and PLSs. Series-22 programmers cover EPROMs and PLEs. Series-22 programmers cover EPROMs and PLEs.

DIGELEC INC.

Europe – Dufourstrasse 116, PO Box 44, CH-8034, Zurich, Switzerland phone: 01 69 38 88; telex: 56913

USA – 7335 East Acoma Drove, Dept 103, Scottsdale, Arizona, Az 85260 phone: (602) 991-7268; telex: 910-550-1301

Capability: UP-803 universal programmers will program all PLDs and allows direct entry of Boolean equations from the keyboard.

ELAN DIGITAL SYSTEMS LTD

UK – 16/20 Kelvin Way, Crawley, West Sussex, RH10 2TS phone: 0293 510448; telex: 877314

Capability: U1000 universal programmer for all PLDs; gang programmers for EPROMs (E series)

FRONTEND LTD

UK – Howell Building, Brunel University, Uxbridge, Middlesex, UB8 3PH phone: 0895 58501

Capability: design entry system which can be interfaced to any other CAE for simulation and logic compilation.

GP INDUSTRIAL ELECTRONICS LTD

UK – Unit E, Huxley Close, Newnham Industrial Estate, Plymouth, PL7 4JN phone: 0752 332961; telex: 42513

Capability: XP640 universal programmer for all PLDs; gang programmers for EPROMs (P series)

KONTRON ELECTRONICS

UK – Blackmoor Lane, Croxley Centre, Watford, Herts, WD1 8XQ phone: 0923 45991; telex: 922012

Germany – Oskar-von-Miller Strasse 1, 8057 Eching phone: 08165 77-0; telex: 526719

USA – 1230 Charleston Road, Mountain View, California, CA 94039 phone: (415) 965-3505; telex: 910-378-5207

Capability: EPP-80 universal programmer for all PLDs, and MPP-80S portable version. UPM module is a single module for the programmers covering all programmable devices.

LOGICAL DEVICES INCORPORATED

UK – Bytron Ltd, High Street, Kirmington, South Humberside, DN39 6YZ phone: 0652 688626; telex: 527339

USA – 1321 NW 65th Place, Fort Lauderdale, Florida, FL 33309 phone: (305) 974-0975; telex: 383142

Capability: ALLPRO software driven programmer uses a personal computer for configuring programming algorithms. PROMPRO-XP universal programmer for PALs and PLEs. Logical also support some standard software packages (PALASM and HELP) under the generic term CAST – Computer Automated Software Tools.

MICROPROSS

UK – Concentrated Programming Ltd, Euro House, 17 Church Street, Rickmansworth, Herts, WD3 1BZ phone: 0923 774418; telex: 918055

France – Parc d'activite des Pres, 5 rue Denis-Papin, 59650 Villeneuve d'Ascq phone: 020 47 90 40; telex: 120611

Capability: ROM5000D programming workstation. Facilities include ASCII keyboard, screen and floppy disk drive for storing programming algorithms and device programs. The system includes PAD (Programmer Aided Design) for assembly and disassembly between Boolean equations and fuse maps.

PERSONAL CAD SYSTEMS INCORPORATED

UK – Petratec Ltd, Glanty House, The Causeway, Egham, Surrey, TW20 9AH phone: 0784 39881; telex: 889204

USA – 1290 Parkmoor Avenue, San Jose, California, CA 95126 phone: (408) 971-1300; telex: 3717199

Capability: PCAD design entry system with interface to CUPL, which is a universal PLD logic input, simulator and compiler. This system, installed on a personal computer and connected to a programmer, provides a full PLD CAE facility.

STAG ELECTRONIC DESIGNS LTD

UK – Stag House, Tewin Court, Welwyn Garden City, Herts, AL7 1AU phone: 07073 32148; telex: 8953451

France – Generim, Z d'A de Courtaboeuf, Avenue de la Baltique, BP 88, 91943 Les Ulis Cedex phone: 0907 78 78; telex: 691700

Germany – Scantec Systemelektronik GmbH, Landshuter Allee 49, D-8 Munchen 19 phone: 089 134093; telex: 5213219

USA – Stag Microsystems, Inc., 528-5 Weddell Drive, Sunnyvale, California, CA 94089 phone: (408) 745-1991; telex: 910-339-9607

Capability: PPZ universal programmer with modules for all PLDs. ZL30/ZL30A programmers dedicated to PALs, PLAs and PLSs. ZL33 gang programmer for erasable PLDs with longer programming times. CUPL is distributed by Stag in Europe and, together with STAGCOM2 and a design entry program, convert a standard personal computer into a complete work-station for PLDs. STAGCOM2 is a software link allowing the PC to take over the handling of the programmer controls, adding editing and display features. Stag also produce a range of gang programmers for EPROMs (PP series).

SUNRISE ELECTRONICS INCORPORATED

UK – Elex Systems, John Scott House, Market Street, Bracknell, Berks, RG12 1JB phone: 0344 52929; telex: 847383

USA – 524 South Vermont Avenue, Glendora, California, CA 91740 phone: (213) 914-1926; telex: 910-584-3847

Capability: Z-1000/Z-1000B universal programmer catering for all PLDs. The Z-2000/Z-2000B offer the same programming capability plus a CP/M based floppy disk system enabling standard software to be run on the same equipment.

TERADYNE INCORPORATED

UK – Teradyne House, Esher Green, Esher, Surrey, KT10 8BN phone: 0372 62199; telex: 929809

France phone: 01 745 17 60

Germany phone: 08940 1961

USA – 321 Harrison Avenue, Boston, Massachusetts, MA 02118

Capability: LASAR Version 6 – includes modelling for PLDs for inclusion in higher level simulation and test analysis.

VALID LOGIC SYSTEMS INCORPORATED

UK – Valid House, 39 Windsor Road, Slough, Berks, SL1 2EE phone: 0753 820101; telex: 847318

USA – 2820 Orchard Park Way, San Jose, California, CA 95134 phone: (408) 945-9400; telex: 3719004

Capability: ValidPLD is an interface between standard JEDEC files and other SCALDsystem design tools, enabling designers to use integrated simulation and test analysis for systems incorporating PLDs.

VALLEY DATA SCIENCES

UK – Data Translation Ltd, The Business Centre, Molly Millar's Lane, Wokingham, Berks, RG11 2QZ phone: 0734 793838; telex: 849862

USA – 2426 Charleston Road, Mountain View, California, CA 94043 phone: (415) 968-2900; telex: 4993481

Capability: 160-series intelligent programmers for all PLDs. These programmers are driven by a personal computer which stores the algorithms on floppy disk. When used in conjunction with VISTA, which is a front-end design tool, and the CUPL compilation and simulation program, they form a complete PLD CAE system.

Appendix 3

Suggested Reading

The following books and papers were useful to me while writing this book. They may also help the reader to gain a greater insight.

CHAPTER 1:

1.1 Ghandi, S.K. (1968) *The Theory and Practice of Microelectronics.* 1st ed. New York: John Wiley.
1.2 Grove, A.S. (1967) *Physics and Technology of Semiconductor Devices.* 1st ed. New York: John Wiley.
1.3 Read, John W. (ed.) (1985) *Gate Arrays* 1st ed. London: Collins.

CHAPTER 2:

2.1 Greene, Bob (1976) *Application of the Intel 2708 8 k Erasable PROM* Santa Clara: Intel Corporation.
2.2 Nelmes, Guy (1985) 'The technology of a 1Mbit CMOS EPROM', *New Electronics.* **18**, 22, 70–74.
2.3 Data I/O Corp. (1978) *How to Survive in the Programming Jungle* Issaquah: Data I/O Corporation.
2.4 Hills, John (1979) 'Higher density goals are foreseeable', *Electronics Weekly*, 25 July 1979.
2.5 Mullard Ltd (1979) *A 16 k PROM – Its Design and Application.* Mullard Technical Note 111, TP1705.

CHAPTER 3:

3.1 Monolithic Memories Incorporated (1984) *PAL/PLE Programmable Logic Handbook* 4th ed. Santa Clara: Monolithic Memories Incorporated.
3.2 Signetics Corp. (1984) *Latches & Flip-Flops with 82S153* Application Note AN14, Sunnyvale: Signetics Corp.

CHAPTERS 4 – 8:

4.1 Bolton, M.J.P. (1985) 'Designing with programmable logic', *Proceedings I.E.E.* 132 E and I, No. 2, March/April 1985, pp 73–85.

4.2 Signetics Corp. (1983) *What is Integrated Fuse Logic?* Application Note AN009, Sunnyvale: Signetics Corp.

4.3 Righter, Bill and Sur, Dean (1986) 'EEPROM technology revitalises PLAs', *Electronic Product Design*, February 1986, 41–47.

The Handbooks and Data sheets of:
Altera Corporation
Advanced Micro Devices
Harris Corporation

Intel Corporation
International CMOS Technology, Inc.
Lattice Semiconductor Corp.
Monolithic Memories, Inc.
Signetics Corporation
Sprague Solid State
VLSI Technology, Inc.

Index